EXPLORING THE LAND AND ROCKS OF SOUTHERN ILLINOIS

A Geological Guide

Stanley E. Harris, Jr.,
C. William Horrell
and
Daniel Irwin

Illustrated
with 191 Maps,
Drawings,
and Photographs

SOUTHERN ILLINOIS
UNIVERSITY PRESS
Carbondale and Edwardsville

Feffer & Simons, Inc.
London and Amsterdam

Library of Congress Cataloging in Publication Data

Harris, Stanley Edwards, 1918–
Exploring the land and rocks of southern Illinois.

Includes bibliographies and index.
1. Geology—Illinois—Guide-books. I. Horrell, C. William, joint author. II. Irwin, Daniel, 1930—joint author. III. Title.
QE106.S58H37 557.73 76-51382
ISBN 0–8093–0740–5
ISBN 0–8093–0769–3 pbk.

Printed in the United States of America
Designed by Gary Gore

Contents

Illustrations

Figures

Charts

Introduction

Aesthetic appreciation of landscape must be part of man's heritage as revealed by the oldest paintings and literature. Early explorers of our nation extolled its beauty and Congress set aside the National Parks for their magnificent scenery. Yet it is not necessary to go far from home to find scenic beauty or to encounter curious features of the rocks and landscape which stimulate intellectual wonderment.

Southern Illinois has beautiful scenery and interesting landforms which attract thousands of nature lovers, picnickers and campers, hunters and fishermen. Good roads and leisure time make every nook and corner of the land accessible. Many scenic spots, such as Garden of the Gods, have become reachable by good roads and paths. The grid is so extensive that representative examples of our landscape could be glimpsed in a day's drive.

Hiking trails lead into wilderness and abandoned areas, and excellent large-scale maps and aerial photographs are readily available for guidance in cross-country walking. Many hikers carry guides to the identification of birds, trees, flowers, and so forth, but guides to the landscape are not so generally used. Most guidebooks attempt to cover a multitude of landforms, from seashore to glaciated mountains. There is too little space devoted to features of plains and hills to be very helpful in our region.

Our purpose in this volume is to describe and illustrate the features, rocks and sediments, which make up our particular landscape. Most of these features are not unique to any one locality, even though they may be beautifully displayed and even set aside in parks, nature preserves, and recreation areas. Such features as cliffs, caves, and canyons may be encountered in hundreds or thousands of unsuspected or little-visited places. Further, the familiar streams and flat upland plains can be very interesting once one is introduced to the detail of their character. The land is not static; dynamic earth processes change with the seasons and after a few years (or a single great storm) cause significant alterations.

Most people who appreciate the scenic also revel in the beauty of nature's detail. The scrolls, involutions, and colors in the sandstone cliffs at Fountain Bluff are enhanced by lichen and moss clinging to the surface, and larger plants whose roots penetrate joints and bedding planes. They wonder how the scrolls were formed and how long it took. Earth scientists still wonder about such things, too, but observation and experiment gradually give insight to nature's processes. We have tried by word and illustration to direct attention to features and processes, large and small which compose the landscape, in the belief that recognition and understanding will enhance the aesthetic appreciation.

The book is divided into four parts. The first describes the physical characteristics of the natural divisions of Southern Illinois and provides information about the climate and the availability and use of maps. The next two describe the features of the landscapes; and the last, the nature of sediments and rocks found in our region.

Each part of the book is divided into sections which discuss the features of the land and rocks. The focus is not on particular places but on particular aspects of the landscape and the materials which compose it. Beneath pertinent section headings are the names (with scale) of representative topographic maps. These maps are available for purchase from the United States Geological Survey, the Illinois Geological Survey, and a few specialty stores such as the University Bookstore on the Southern Illinois University campus at Carbondale (see addresses Appendix A).

The descriptions and illustrations in this book will help the reader recognize many features and associations which might otherwise have been overlooked. Explanations are brief and not intended to give full scientific data. At the end of most sections a few suggested readings are listed which give more complete explanations. Where no reading is listed, one of the general references following the section on Earth Processes may be used. If your curiosity is sufficiently aroused, an appropriate modern textbook on physical geology, geomorphology, petrology, mineralogy, or paleontology would be useful. The reports of the Illinois Geological Survey are available free of charge; those out of print can be borrowed from university and many college and city libraries.

A small index map beside the heading of many sections of the book indicates the distribution of the feature. Where no map appears, the distribution is general.

Topographic maps are distributed through the book to illustrate landscape relationships. These are copied segments of United States Geological Survey maps. The caption identifies the source map which can be purchased at the places listed in Appendix A. The contour lines depict the aspect of the land. Along each line the elevation above sea level is constant. The closer the lines are to one another, the steeper the slope; and the more lines there are, the greater the height. The vertical intervals between contours are ten or twenty feet, depending on the map.

At the end of most sections a paragraph headed "Things to do" gives suggestions for observations and experiments which might serve to give better understanding of the features described. These could be especially useful for school projects.

Our book does not attempt to summarize the geologic story of Illinois nor does it systematically review the different natural environmental conditions which produced the rocks, sediments, and landscape features. A short section on Geologic Time in part one attempts to put into perspective the great span of time during which our region has evolved. We know about events of the past only as they are recorded in the rocks. Sedimentary rocks have been laid down layer on layer with the oldest on the bottom. The column of sedimentary rocks seen in exposures or encountered while drilling wells gives clues to the changing environments. The total thickness as seen in the deepest part of the Illinois Basin has over twelve thousand feet of sedimentary rocks. The various rock units have been given names taken from the localities where they were first described at the surface. The chart in Appendix C lists these formation names in order, with the youngest at the top. We include this because the names often appear in newspaper articles reporting a new oil pool or a coal mine or quarry. The chart will indicate the geologic age of the formation and the relative position in the rock column.

The photographs and diagrams were made specifically to illustrate Southern Illinois rocks and topographic features. The aerial photographs give a perspective which shows synoptic relationships while ground photographs give more detail.

Our book is a guide to the rocks and the topographic features which you will see in Southern Illinois. We mention some places to visit but the joy will be in your own discovery and recognition.

The authors wish to acknowledge the assistance provided by Southern Illinois University Graduate School and the Geology Department in the research and preparation of the photographs and the typing of the manuscript.

All of the topographic maps are copied from parts of quadrangle sheets published by the United States Geological Survey. Several maps, as indicated, were taken from publications of the Illinois State Geological Survey. We are especially pleased with the Earth Resources Technical Satellite (ERTS) imagery of

Southern Illinois, supplied by the Earth Resources Observation Satellite (EROS) Data Center.

We have benefited from the ideas and suggestions of colleagues and friends who have taken an interest in our project. The "everyday" questions about the earth, received over the past years from students and field-trip participants, provided the basis of the book. Our special thanks go to Ann Pipkins, Wanda Rudolph, Dotti Davis, and Karen Schmit for their help in its final preparation.

STANLEY E. HARRIS, JR.
C. WILLIAM HORRELL
DANIEL IRWIN

Southern Illinois University-Carbondale
March 15, 1976

Geology and Landscape

1. Illinois River Bottomlands
2. Western Forest Prairie
3. Middle Mississippi Border
4. Southern Till Plain
5. Wabash Border
6. Ozark
   a northern
   b central
   c southern
7. Lower Mississippi River Bottomlands
8. Shawnee Hills
9. Coastal Plain

MISSISSIPPI
RIVER
WABASH
RIVER
OHIO
RIVER
SCALE IN MILES
0
20

1. The Natural Divisions of Southern Illinois.

Geologic Framework

Where in Southern Illinois are the cliffs, shelter bluffs, and high lookout places which publicity brochures tell about? Where are the caves or waterfalls or old glacial moraines? Where does one find glacial boulders or agates or fluorspar?

The best guide is a regional map showing distributions and patterns of rock formations and landforms. These are reflections of the geologic framework and the active processes which have left their imprint on the land. There exist no firsthand records of the slow changes and events of the past. Rather, the earth scientist started with observations of the landform, the rocks, sediments, and soils, and gradually reconstructed the story of what must have happened in the last half billion years to produce the landscape we live in. Our story is incomplete. There is much left to discover and interpret. We hope this book will help sharpen your powers of observation so you will see a great deal in the rocks and the landscape. Then you will wonder how they came to be that way and soon you will be making interpretations of your own.

The natural divisions map (Fig. 1) is very useful because each division contains a certain unity of geologic, surface and biologic features. The sections describing characteristics of each division indicate the types of features which one might expect to see. The bedrock geology map (Fig. 2) shows boundaries, also bands within which certain rock formations and surface features occur, that is, coal strip mines and the margin of glacial deposits.

Landscape

The landscape is a product of erosional sculpturing of a basic framework of sedimentary rocks overlain in part by a surficial veneer of later sedimentary deposits. Visualize the Midwest as a region resting on a pile of layered rocks only a few thousand feet thick but stretching a thousand miles westward from the Appalachians and six hundred miles in width from north to south. The rocks are of sandstone, shale, limestone, and dolomite. Some of the latter are flinty or contain flint bands. Thin, but remarkably persistent coal beds occur in the upper part of the sequence. The beds of rock themselves are not continuous. Some layers merely pinch out (Fig. 3), others change laterally to a finer or coarser texture. Each variety of rock has a wide range of detailed properties and compositions.

In the last billion years no mountain building has disturbed the rock layers of the Midwest. In a quarry or bluff they seem to be nearly horizontal. Regionally they are gently flexed in the form of broad basins and arches. The upper layers of rock were eroded from the tops of the arches but still persist in the center of the basins where they extend far below sea level (Fig. 4).

The Shawnee Hills Division

The backbone of Southern Illinois is a continuous ridge extending from Battery Rock overlooking the Ohio River to Horseshoe Bluff overlooking the Mississippi Valley. The ridge rises westward beyond Big Muddy Valley into Randolph County where it dies out. Physiographically we call it a "cuesta," a long asymmetrical ridge with a steep slope on the south side and a long gentle backslope inclining toward the north. The highest elevations and deepest canyons are found along this ridge. Here is our loveliest scenery. The crest is held up by resistant massive sandstones exposed

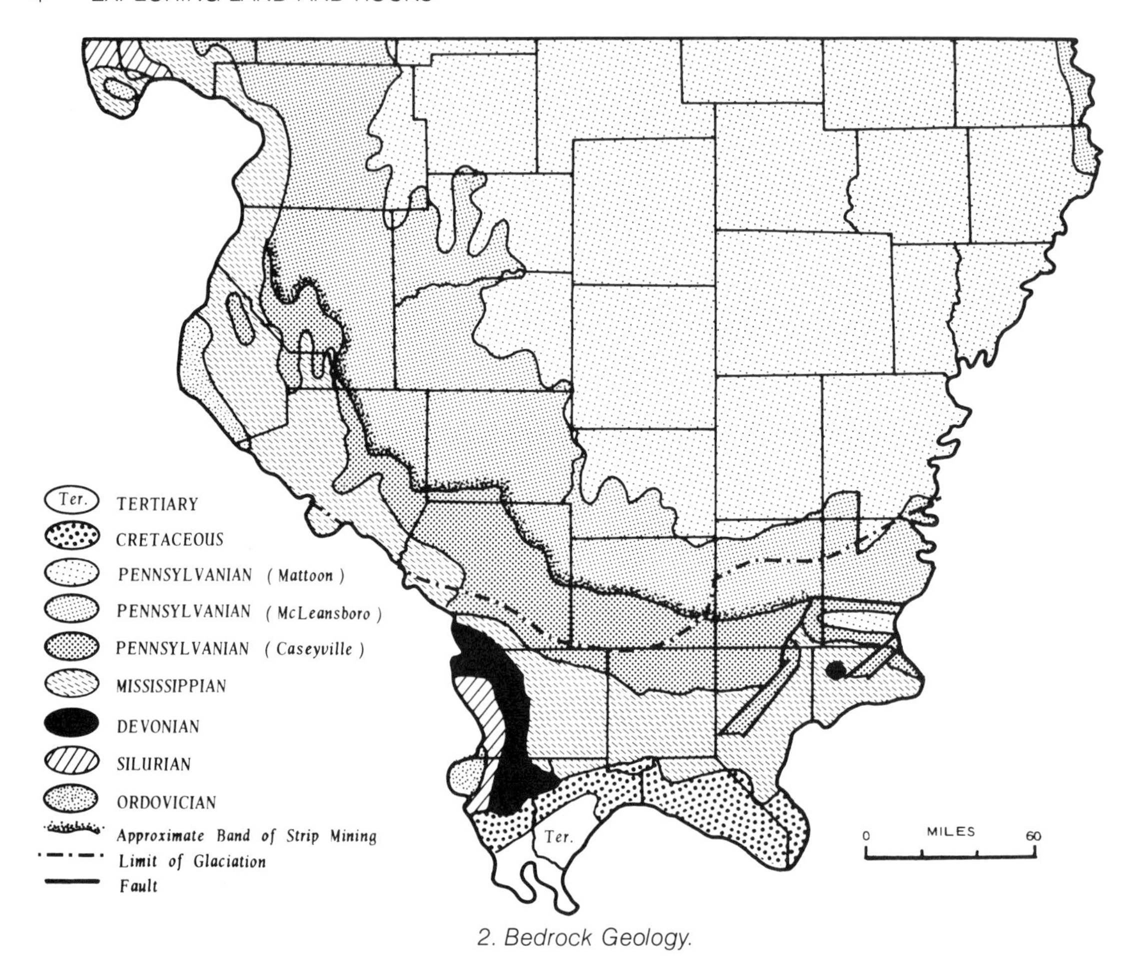

2. Bedrock Geology.

in cliffs along the escarpment. In the other direction the rocks dip gently toward the north at a very low angle producing the gentle backslope. Actually the cuesta is compounded by several repetitions of rise and step (Fig. *5*), each crest held up by massive sandstone. The dominant cuesta is produced by the Caseyville sandstones of Pennsylvanian age. The rocks are best exposed at cliff sites such as Garden of the Gods, Draper's Bluff, Ferne Clyffe, and in the deep road cuts of Interstate 57 and Interstate 24. The natural divisions map (Fig. *1*) clearly shows this topographic feature along the line marking the southern boundary of the Pennsylvanian rocks.

South of the major east-west cuesta a series of smaller parallel cuestas trend a little more southeastward. These ridgetops have sandstone beneath them also. Their elevation is not quite so high nor is the rock so continuously massive. The streams of the region tend to flow parallel to the ridges but now and again turn nearly at right angles and cut through the ridges with narrow watergaps (Fig. *6*). Where the streams lie parallel to the ridges, they occupy the outcrop of the more easily eroded shale and limestone formations. The floodplains are broad and are not the product of the streams themselves. Rather, the sediment is a backwater fill many feet thick caused by glacial outwash brought two hundred miles down the major rivers beyond the Wisconsin glacial margin.

3. *Fine-grained sandstone beds at the spillway of Kinkaid Lake, Jackson County. Note that the many beds change thickness or disappear even within the space of the photograph.*

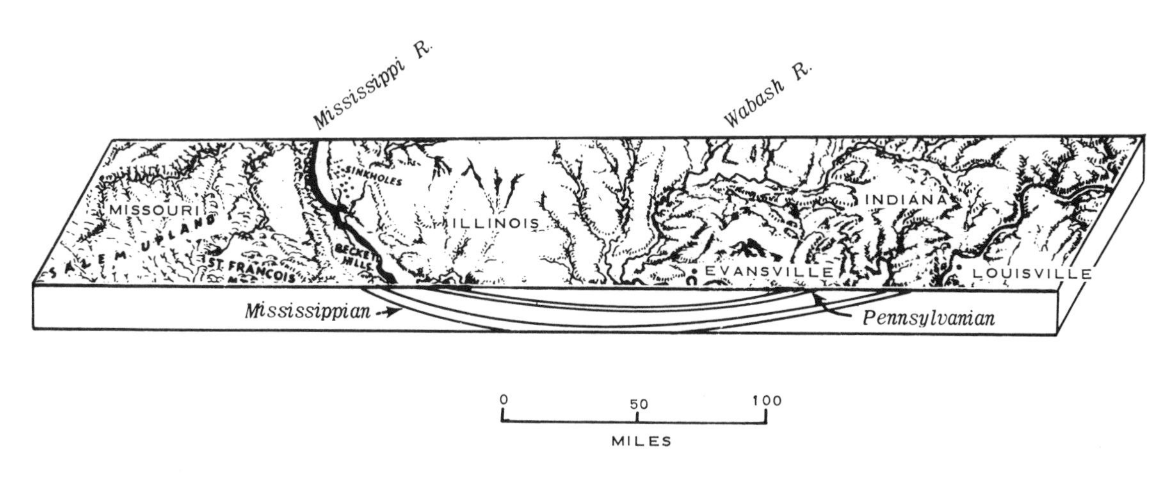

4. *Diagram and cross section showing the Illinois structural basin between the Ozark dome on the west and the Cincinnati arch on the east.*

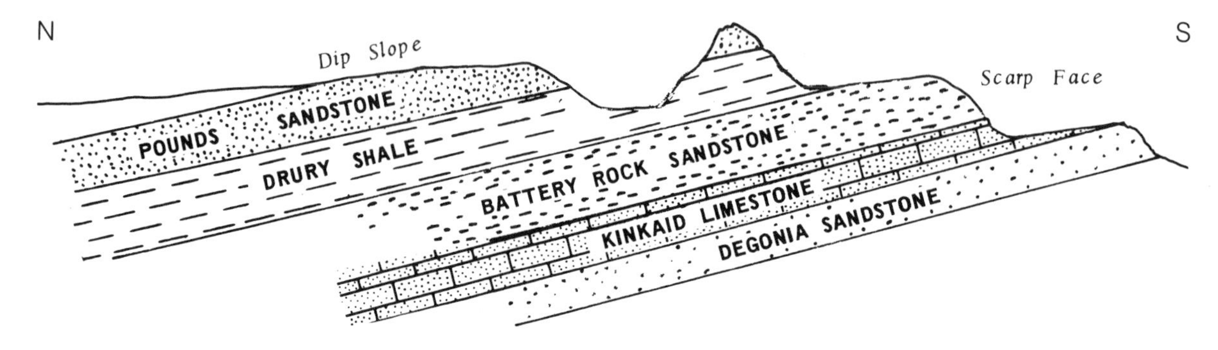

5. Cross-section Profile of Cuesta South of Carbondale.

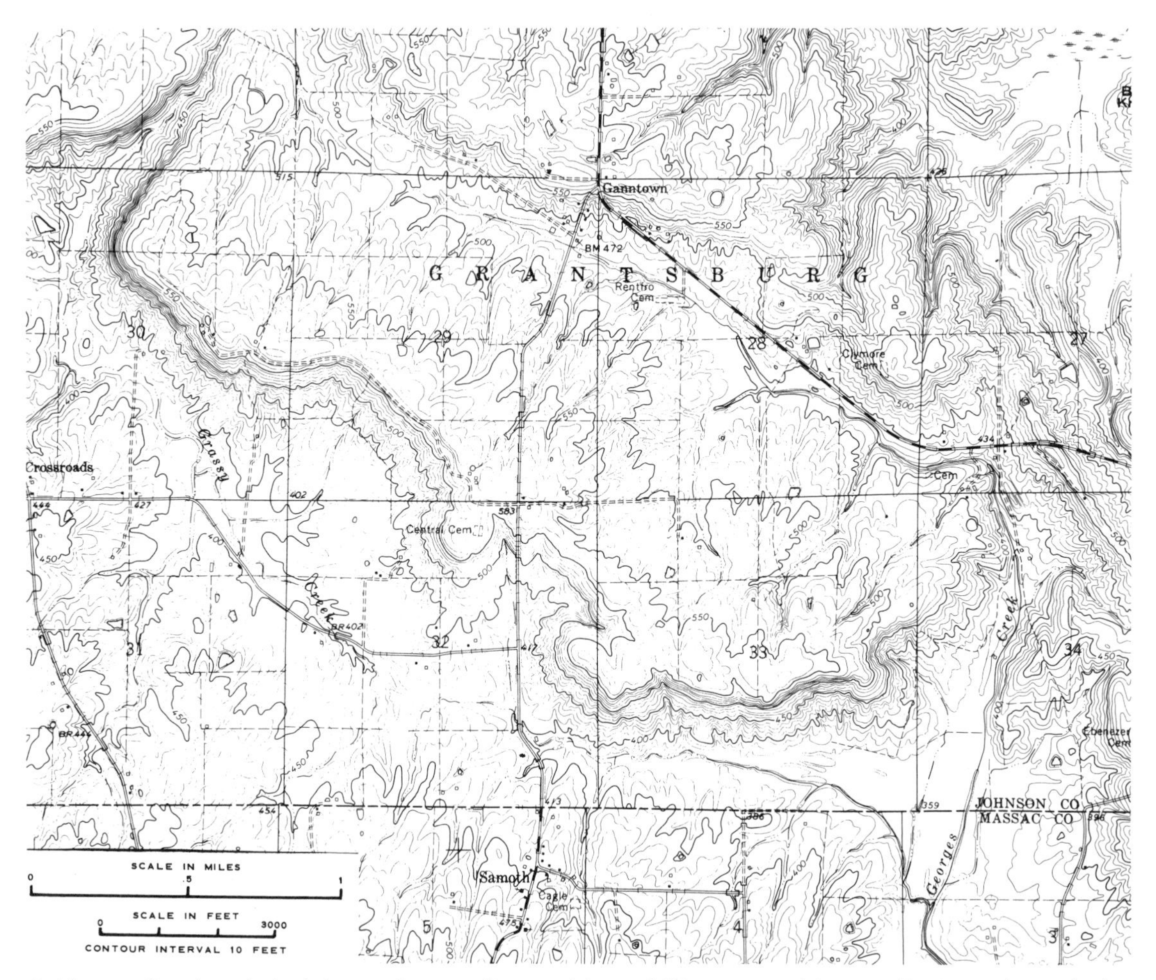

6. Mermet Quadrangle in Johnson County. Cuesta ridges of Chester sandstones with steep slopes facing southwest and dip slope to the northeast. Stream valleys of this part of the Shawnee Hills trend along the outcrop of the more easily eroded limestone and shale, but most do cross through the ridges in narrow water gaps. Note Georges Creek in the southeast.

7. Little Muddy River bottomland in Jackson and Franklin counties. Open areas in foreground are the Campbell Lakes in the wooded bottomlands that are subject to frequent floods.

Southern Till Plains Division

The Illinois coal basin lies north of the Shawnee Hills. Relief is much less and the underlying bedrock is dominantly shale and thin sandstones. Flat and rolling uplands characterize the region. Cultivated farmlands were once great tallgrass prairies. Most of the larger stream valleys are wide and subject to frequent floods. The aerial photograph (Fig. 7) shows most wooded bottomlands. The many tributaries near each major stream have dissected the land, producing steep slopes and varied relief not suitable for cultivation. Wooded hilly areas with lovely valleys and low cliffs are found throughout the Mt. Vernon Hill Country Section especially in Jefferson and Hamilton counties. Northward in the Effingham Plain Section along U.S. Route 40 the land is flat or rolling.

The edge of the commercial coal beds is clearly marked today by the discontinuous band of strip mines extending from St. Clair County in an arc to Gallatin County. The coals dip eastward or northward, sloping deeper and deeper into the basin to its deepest part in White and Wabash counties where they lie twelve hundred feet beneath the surface. The rocks above the coal are mostly shales with some interbedded, thin and generally impure limestones, and sandstones much less massive than those which form the Shawnee Hills. The outcrop belts of two limestone units are shown in Figure *2*.

Glacial till deposits (Fig. *8*) cover the bedrock throughout the plains division. The southern boundary marked on the natural divisions map (Fig. *1*) is the approximate limit of advance of the Illinoian ice sheet. The glaciers did push their way into the valleys and up the backslope of the Shawnee Hills, and into the valley of the Mississippi River in western Jackson County.

Yet till is not the uppermost deposit. At the time of the next glaciation, the Wisconsonian, summer meltwaters carried vast quantities of sediment down the rivers. Each winter the freezing cold shut off the river flow and loose sediments were left on barren river flats. Winds picked up the fine-grained material; clay sizes must have been blown far to the east and south while sand sizes merely produced dunes. The silt particles were swept out of the valley bottoms but dropped into the woods, lakes, and grassy plains. This silt mantles most of Southern Illinois so that most of our soils have developed in it. It becomes sticky mud when wet but technically contains little clay. We call this deposit "loess"; the name is of German derivation and is applied to similar material of the Rhine valley. Vast windblown loess deposits also cover northeastern China and Manchuria. At the Mississippi bluffs near-vertical road cuts expose twenty feet or more of the buff-colored loess (Fig. *9*).

Ozark Division

Along the western side of the area a sliver of the Ozarks is responsible for the great limestone bluffs marking the edge of the Mississippi Valley (Fig. *10*) and the sinkhole region of Monroe County with its caves and springs. Pine Hills and the flint-covered hills at Bald Knob and Atwood Ridge also belong to the Ozarks.

8. Very stony glacial till containing some pebbles of crystalline, igneous, and metamorphic rocks brought from Canada.

9. Road cut on Chester bypass, Randolph County. The silty loess encloses snails and peculiarly shaped concretions of calcium carbonate.

10. Bluff of Bailey Limestone at Pine Hills in the Ozark Division.

The southern section of the Ozark Division has high relief and very steep slopes, mostly wooded. It does not contain a single town. Except for the fact that it has been timbered repeatedly it is a wilderness area. Small farms and residences occupy the alluviated valley floors. Relatively few ridgetops are broad enough to be cultivated.

Bottomland Divisions

Enclosing Southern Illinois on three sides are the great rivers, Mississippi, Ohio, and Wabash. These have broad floodplains, and related bottomlands extend up their tributaries. The Mississippi is confined by high limestone bluffs. The Ohio valley is also trenchlike with some high bluffs but generally a more subdued valley side; the floodplain lies largely on the Kentucky side of the river. The Wabash Valley is very wide with no confining valley wall. It has broad flat terraces and the uplands rise gently above them (Fig. *11*).

Coastal Plain Division

At the extreme south is a small segment of coastal plain including the bottomlands of Cache-Bay creeks and the Cretaceous Hills Section bordering the Ohio River. The bottomlands are similar to the great river bottomlands except that they contain no river and their

11. A bend of the Wabash River at New Haven, White County, looking south toward Shawneetown Hills. The terrace in foreground is flooded occasionally and the broad flat surfaces may drain slowly except where sandy. No bluffs mark the edge of the valley.

natural vegetation has southern affinities. The hills are formed from clays, sands, and gravels. Slopes are steep and subject to rapid erosion and earth slides. Flat terraces and broad bottomlands extend into the hilly areas (Fig. *12*).

SUGGESTED READING

Hunt, Charles B. 1974. *Natural Regions of the United States and Canada*. San Francisco: W. H. Freeman. 725 pp.

Illinois Nature Preserves Commission. 1973. *The Natural Divisions of Illinois*. Comprehensive Plan for the Nature Preserves System Part 2. Springfield. 32 pp.

Leighton, M.M.; Eckblaw, G.E.; and Horberg, Leland. 1948. *Physiographic Divisions of Illinois*. Urbana: Illinois State Geological Survey, Reports of Investigations no. 129. 19 pp.

MacClintock, Paul. 1929. *Physiographic Division of the Area Covered by the Illinoian Drift-Sheet in Southern Illinois*. Urbana: Illinois State Geological Survey, Reports of Investigations no. 19. 57 pp., 27 figs.

Thornbury, William D. 1965. *Regional Geomorphology of the United States*. New York: Wiley, 609 pp.

Willman, H.B. et al. 1968. *Bibliography and Index of Illinois Geology Through 1965*. Urbana: Illinois State Geological Survey, Bulletin no. 92. 373 pp.

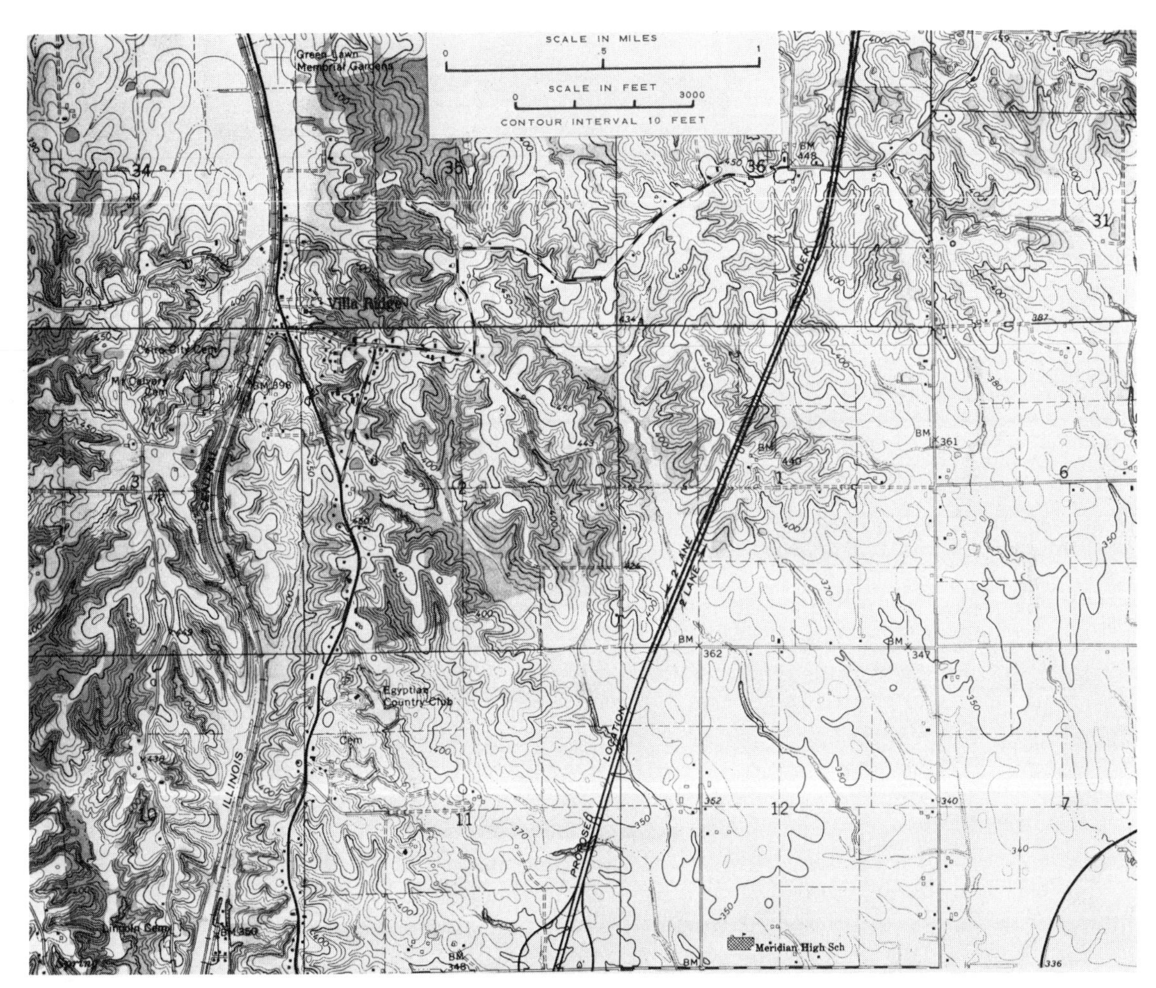

12. Pulaski topographic map, a part of the Coastal Plain Division. The wide, flat terraces next to hilly areas with a dense network of small valleys are typical. Severe erosion and deep gullying into sands and clays are serious hazards.

Earth Processes

The geologic framework of a region sets the stage. Differing rock properties, their arrangement and structure, are the raw materials. Geologic events and their sequence through time, especially as they affect elevation of the land and distance from the sea, help determine what the landscape will be like. Ultimately, though, it is the dynamic surface processes which sculpt the land and determine its form and aspect.

When exploring the land, observe the processes at work as they gradually erode here and deposit there or chemically alter soil and rock. The land is beautiful and its differing seasonal garb makes it ever changing. Watch its gradual modification and learn to recognize evidences of change from the past and learn to anticipate and predict changes yet to come. Even though we all know that erosion and deposition go on all the time, perhaps we are not aware of the considerable alterations that can take place in a person's lifetime or even in a year's time.

The climate of our region determines the surface processes at work. Continually changing intensity levels are just as important as the variety of processes. The sun's energy determines the temperature of the ground and, indirectly, of the air. Greatest radiant energy is received when the sun's rays fall at right angles to the surface. On an average this would be in June, though a south-facing hillslope would receive more energy than a north-facing slope. Hence, there is more evaporation on the south-facing slope, more repeated wetting and drying in wet months, more repeated freezing and thawing in winter months. Vegetation reflects this and so does soil.

The movements of air masses and the weather which accompanies them is of great importance. These determine temperature, precipitation, moisture, and their patterns. Cycles of weather or intervals of drought or wetness are of long-term importance. These determine the specific dynamic processes of running water, groundwater, chemical and physical alterations, and contribute to the instability of some land surfaces. Furthermore, we have the clear evidence that in the immediate geologic past the climate was cold enough to allow continental glaciers to reach across the central lowlands and profoundly affect the landscape.

In this section we will do little more than enumerate the processes at work and illustrate the direct impact they are making. Raindrop impact raises much soil into the moving water film. Drip from a cliff splatters more fine material downslope than upslope. Surface runoff carries soil particles downslope depositing them in the next puddle or at the foot of a gully in a fanlike deposit (Fig. *13*). Concentrated streamflow carries small and large particles especially in floodwaters. The size of the boulders scooted against a tree or onto a bar at the end of a bare rock chute makes one wonder at the force possessed by moving water (Fig. *14*).

Many types of mass-wasting bring material downslope. Gravity is the direct cause (Fig. *15*). Imbalance comes about by erosion (or undercutting by man), by adding weight to, or reducing friction within, a loosened mass of rock. Landslides, slumps, slow creep, wedging of roots (Fig. *16*) move incredibly large masses of material from a slope into a stream channel. The combination of gravity and the force of running water gradually deepen and widen the valleys (Fig. *17*).

Rock and soil are slowly altered by the many chemical and physical processes which we include under weathering. Percolation of water through the soil dissolves certain elements making them available to plants but also carrying them to greater depths or perhaps into streams by way of seeps and springs. Underground water percolating through the crevices

13. Surface runoff leaves fanlike deposit of soil particles at foot of gully.

14. Gravel and bouldery bar in a stream bed in Giant City State Park, Union County. The overlapping (imbricated) arrangement of the large slabby blocks show that they have been scooted along the stream channel floor during high water, finally coming to rest against another boulder or obstruction. Note smooth rocky channel upstream across which boulders have slid.

15. Rock fall of slabby siltstone, undercut by a tributary of Lick Creek in Union County.

16. Tree root has grown along bedding plane of a collapsed block of sandstone in Ferne Clyffe State Park, wedging apart the upper layer.

17. Slide blocks on the valley slope and at the edge of the stream channel (middle ground) near Makanda, Jackson County. Once gravity has brought them to the channel they are subject to impact and breaking by other boulders transported by floodwaters. Eventually they are chipped down to transportable size.

18. Massive, fine-grained limestone on Hutchins Creek, Union County. Note the rounded surfaces and dissolved channels and openings.

of limestone creates caves and cavities (Fig. *18*). Extensive networks may capture all surface water so that no surface streams can persist. Collapsed cavities cause sinkholes at the ground surface. Subsidence of the land into abandoned mines produces similar features.

Lakes are dynamic places. Waves erode the shores, producing cliffs, and currents carry the eroded product onto the lake floor. This together with sediment brought by surface runoff gradually reduces the storage capacity of the lake.

Organisms constantly affect the land. The vegetation plays a great role in reducing immediate surface runoff but also creates humic acids which facilitate the weathering and solution of soil particles and underlying bedrock. Microorganisms in the soil are of utmost importance in the operation of the nitrogen-carbon dioxide-oxygen cycle. Earthworms and other burrowing animals turn the soil, bringing up new materials to be weathered and to release its nutrients while allowing organic matter to fall down into the opening. Grazing animals often create paths down which water flows, perhaps creating gullies.

At this point in geologic history perhaps man has become the geatest force for change in the landscape. While all other processes have their specific role and locus of impact, man's activities are ubiquitous. Ancient man had modest impact, although deforestation, grazing, and cultivation led to destruction of soil and the encroachment of desert. Today as an earthmover, alterer of rivers, introducer and concentrator of active chemicals, man may have become the most profound of all geologic processes.

SUGGESTED READING

Adams, G. F., and Wyckoff, J. 1971. *Landforms: A Golden Science Guide*. New York: Golden Press.

Bloom, A. L. 1969. *The Surface of the Earth*. Englewood Cliffs, N.J.: Prentice-Hall. 152 pp.

Byrne, Frank. 1974. *Earth and Man*. Dubuque, Iowa: William C. Brown Co. 765 pp.

Dury, G. H. 1959. *The Face of the Earth*. Baltimore: Penguin Books. 223 pp.

Hamblin, W. K. 1975. *The Earth's Dynamic Systems*. Minneapolis: Burgess Publishing Co. 578 pp.

Money, D. C. 1970. *The Earth's Surface: Physical Geography in Colour*. London, Eng.: Evans Brothers. 168 pp.

Sawyer, K. E. 1970. *Landscape Studies*. London, Eng.: Edward Arnold. 149 pp.

Sharpe, C. F. Steward. 1938. *Landslides and Related Phenomena*. New York: Columbia Univ. Press. 137 pp.

Shelton, J. S. 1966. *Geology Illustrated*. San Francisco and London: W. H. Freeman. 434 pp.

Stokes, W. L., and Judson, S. 1968. *Introduction to Geology*. Englewood Cliffs, N.J.: Prentice-Hall. 530 pp.

Strahler, Arthur, N. 1971. *The Earth Sciences*. 2d ed. New York: Harper & Row. 824 pp.

Tuttle, S.C. 1975. *Landforms and the Landscape*. 2d ed. Dubuque, Iowa: William C. Brown Co. 152 pp.

Geologic Time

The Rock Column

How old are the rocks of Southern Illinois? When did the coal form? How long did it take to accumulate the thirty feet or so of vegetation debris necessary to form one foot of bituminous coal? Such questions about time are fundamental, yet geologists can give only partial, perhaps equivocal answers.

The geologic time and rock chart (Chart 1) gives dates in millions of years derived from radiometric dating methods. For a description of the method and principles involved, please refer to sources such as the references listed under Suggested Reading. The $carbon^{14}$ to $carbon^{12}$ ratio method is perhaps most familiar because it has been applied to the dating of human artifacts and wood, charcoal and carbonate shells in thousands of deposits. However, the $carbon^{14}$ has a half life (half has disappeared) of only 5,570 years. After about 33,000 years the tiny remaining quantities make ratios somewhat questionable. The Illinoian glacial till in Southern Illinois is so old that no C^{14} is present in the enclosed wood, see Chart 2.

The geologic time scale is derived from the column of rocks, especially the fossil-bearing sedimentary rocks. The geologic column is based on: 1) the principle of superposition—that the oldest sedimentary layer is on the bottom; and 2) that the enclosed fossil remains represent a unified evolutionary system stemming from earliest geologic history. Within a region such as ours this is easily shown, though the correlation of our rock column with the Appalachians or Asia is more complex. When visiting a quarry, note the sequence of layers. Some layers may be readily traceable throughout, others pinch out. A quarry nearby will show similarities, but probably not an exact duplication. Each layer of rock represents an event in the past brought about by a certain set of environmental conditions. Without the rock there is no record of the event. We can recognize places where erosion took place; sometimes erosional conditions persisted long enough to remove part of the previous record. The imprint of the erosional processes on the affected rock is itself a record of an event.

Some dates are thought provoking. Glaciers existed at Chicago only 12,000 years ago, but the swamps in which vegetation accumulated, ultimately to become our coal, existed some 300 million years ago. In fact the surface materials deposited by glaciers, winds, and rivers are of late origin but the bedrock is old. The period from Pennsylvanian to Pleistocene is without record over most of our area. During that time dinosaurs roamed the earth and we assume that Illinois was a land inhabited abundantly by plant and animal life. Yet here decomposition and erosion were dominant processes, leaving no deposits to preserve their remains. Cretaceous and early Tertiary sediments are found only at the southern tip of the state. Time is meaningful only in the framework of successive events or as an interval required for changes to occur. The successive layering of sedimentary rocks represents successive events, and the changes in rock types record changes in the local environment. So far, scientists have been unable to determine the length of time required for the accumulation of a given thickness of sediment. We do know that some fine-grained muds accumulate very slowly, while in just one storm some coarse-grained sand and gravel deposits may exceed ten feet.

The time units of the geologic column represent a compilation of the interlinking records from around the world. The rock column represents the sequence of rock units as they would be found when drilling a well. It is generalized because from place to place the units change in thickness or wedge out or

1. GEOLOGIC TIME AND ROCK CHART FOR SOUTHERN ILLINOIS *

Era	Million Years Ago	Periods	Epochs
		Quaternary	*Pleistocene*
Cenozoic	*2*		*Pliocene*
	12	*Tertiary*	*Miocene*
	25		*Oligocene*
	40		*Eocene*
	60		*Paleocene*
	70		
Mesozoic		*Cretaceous*	
	135		
		Jurassic	
	180		
		Triassic	
	225		
		Permian	
	270		
		Pennsylvanian	
		Mississippian	*Chesterian*
			Valmeyeran
			Kinderhookian
	345		
Paleozoic		*Devonian*	
	405	*Silurian*	
	425		
		Ordovician	
	500		
		Cambrian	
	600		

*** The appendix contains a more complete column with formation names.**

Rock Types	Conditions and Events
Wind blown loess	*soil formation*
silt, sand and clay	*river and lake deposits*
till, sand and gravel	*glaciation*
chert gravels	
absent	
absent	*erosion*
clay	*coastal plain*
sand	
and gravel	*deposits*
absent	*erosional processes*
absent	*crustal warping*
absent	
shale, sandstone, limestone, coal	*coastal plain swamps*
	deltas shallow marine
limestone and sandstone with shale formations	*deltas and shallow marine*
limestone mainly	*marine*
shale and limestone	*shallow marine*
limestone fossiliferous	*marine*
limestone exposed only in Ozark Division	
limestone exposed only in Ozark Division	
not exposed	

2. PLEISTOCENE TIME CHART

Time	Years Before Present	Event and Sediment
Holocene Stage		*Formation of modern soils and present—day topography*
	7,000	
Wisconsinan Glacial Stage		*Five ice advances and retreats* *Did not reach Southern Illinois* *Sand and gravel trains on Wabash, Kaskaskia, Mississippi and Ohio Rivers* *Backwater lake sediments on Big Muddy, Saline systems* *Repeated deposits of windblown loess (silt)*
	75, 000	
Sangamon Interglacial		*Soil formation, erosion*
Illinoian Glacial Stage		*At least two glacial advances leaving till and some stratified materials, partly buried older topography* *Ice front rode up into Shawnee Hills* *Briefly entered Mississippi Valley*
Yarmouthian Interglacial		*Soil formation and erosion*
Kansan Glacial *		*Glaciers probably reached southwestern Illinois but evidence is scanty*
Aftonian Interglacial *		*Soil formation*
Nebraskan Glacial *	*1 Million (?)*	

**Not readily recognizable*

grade laterally into rock with different characteristics. Geologic periods are not of equal length, neither are rock formations of equal thickness. Formations are rock units with a distinctive identity which have been mapped over a considerable area. Unfortunately, as in most classifications, the same names have been used in different ways, so there may not be complete equivalence from one publication to another. The chart in Appendix C gives the usage of the Illinois Geological Survey as of spring 1975.

Graphic representation of the rock sequence is given in many reports and field guides. It is customary to arrange the column as it is seen in nature with the oldest at the bottom.

Things to do. Make careful representations of the sequence of rocks in quarries, road cuts, cliffs, and so forth, in a local region. Collect samples and fossils from each unit. Try to match them. Obtain reports and guidebooks from the Illinois Geological Survey, and help from an earth science teacher, local geologist, or staff members of a college Geology Department.

SUGGESTED READING

Radiometric dating

Deevey, E. S., Jr. 1952. Radiocarbon Dating. *Scientific American* 186:24–28. (Offprint no. 872. San Francisco: W. H. Freeman.)

Eicher, D. L. 1968. *Geologic Time*. Foundations of Earth Science Series. Englewood Cliffs, N.J.: Prentice-Hall. 150 pp.

See textbooks in physical or historical geology.

Geologic column

Bradbury, J. C., and Atherton, E. 1965. *The Precambrian Basement in Illinois*. Urbana: Illinois State Geological Survey, Circular no. 382. 13 pp.

Illinois State Geological Survey. Geological Science Field Trip Guide Leaflets available for many localities; new leaflets published every year containing geologic rock column, time chart, and description of the geologic history. Urbana.

Willman, H. B., and Frye, J. C. 1970. *Pleistocene Stratigraphy of Illinois*. Urbana: Illinois State Geological Survey, Bulletin no. 94. 204 pp.

Willman, H. B. et al. 1975. *Handbook of Illinois Stratigraphy*. Urbana: Illinois State Geological Survey, Bulletin no. 95. 261 pp.

Wyckoff, Jerome. 1966. *Rock, Time, and Landforms*. New York: Harper & Row. 372 pp.

See textbooks in principles of geology.

The Uses of Maps

Maps are the keys to discovery of the outdoor world. Today there are hundreds of maps showing locations of parks, historic and scenic places, of distributions of soil types, of mines and quarries, and of vegetation and rock formations. Above all, topographic maps represent land slopes by contour lines and give the location of streams and highways, cliffs and trails, power lines and old roads. Every house, every cemetery, and the sites of many an abandoned farm are shown.

Most maps are readily available and rather inexpensive, although accumulation of detailed maps covering large areas represents a considerable investment. This book includes numerous map excerpts and, at the beginning of many sections, the names of representative maps which show the features described in this section. The principal map sources together with the type of material represented are listed in Appendix A. The maps of Southern Illinois, Figures *1* and *2*, show regional distributions, boundary lines, and zones pointing up special attractions.

The United States Geological Survey quadrangle and United States Forest Service township topographic maps are best for fieldwork or for hiking. A little practice and use bring proficiency in judging slope and distance, the locations of cliffs and perhaps waterfalls; sinkholes and regions where springs and caves might be discovered; west-facing cliff tops where xeric plants might occur or north-facing sites protected from hot sun and high evaporation rates. An index map showing areas covered by each quadrangle map may be obtained free from the Illinois Geological Survey at Urbana. The index map is most helpful and desirable if one plans to order maps.

Make notes on your maps—use a sharp pencil and print neatly. Show the route of your walk, local names, the place where you saw that covey of quail or the pink lady's slipper, the exposure of fossiliferous limestone. Use the wide margin of the map for notes. Perhaps you will wish to use colored pencils or ink to denote features for permanent record. The more you record, the more valuable your map becomes.

Fold field maps face outward in accordian fashion, the way that the *National Geographic* maps and most road maps are folded. A little soil, even mud, won't damage the map, but folding and unfolding will. It can be folded into a convenient size to carry on a clipboard or in a carrying bag. When you order your first topographic map, ask for the booklet describing the legend. Study the map at home and in the field to become familiar with the symbols.

When first using a map, acquaint yourself with the vital information about its construction. Scale is indicated by a ratio such as 1:62,500, meaning that any unit of measure on the map is multiplied by 62,500 for the same distance on the ground. The bar scales are very useful but note the "0" is not at the end of the bar. The left side of "0" has divisions in tenths. If you measure a distance along a road as something more than a mile, place your distance tick at the *right* end, then you can easily read the decimal off the subdivided scale. The edges of the map are oriented north-south and east-west, but the arrow symbols indicate how far your compass needle is off true north. The declination is only a few degrees in Southern Illinois so that adjustment is necessary only if you must be very accurate.

The grid systems shown on a topographic map are used for locating a piece of property or even a point. The civil divisions of Illinois are based on a grid system of townships and ranges divided into square miles. The section lines and section numbers are clearly shown on topographic maps. In addition one or more grids are marked by colored ticks along the margins of the map.

On some maps especially along the bordering rivers, boundary lines askew to the section lines show properties granted by the French (or English) before the American civil land survey was made.

In the original land surveys a N-S meridian and E-W base line were chosen arbitrarily. The diagrams (Fig. *19*) illustrate the manner of construction and the designations attached to the various parts. Each township-range is divided into 36 square miles numbered as shown in the diagram. Most sections were not surveyed accurately so the number of square feet does

SCALE 1:62500

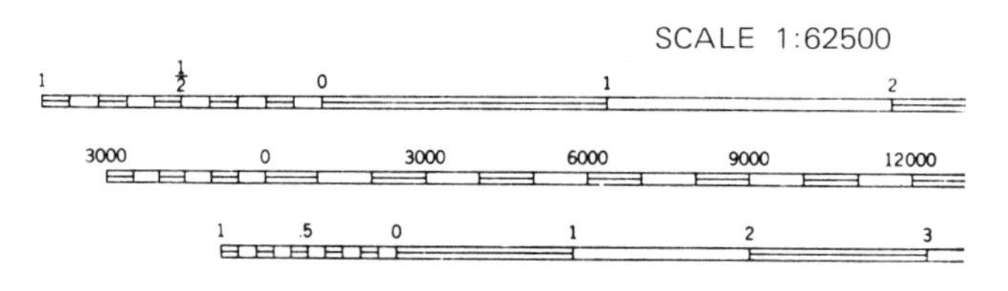

Contour interval 20 feet

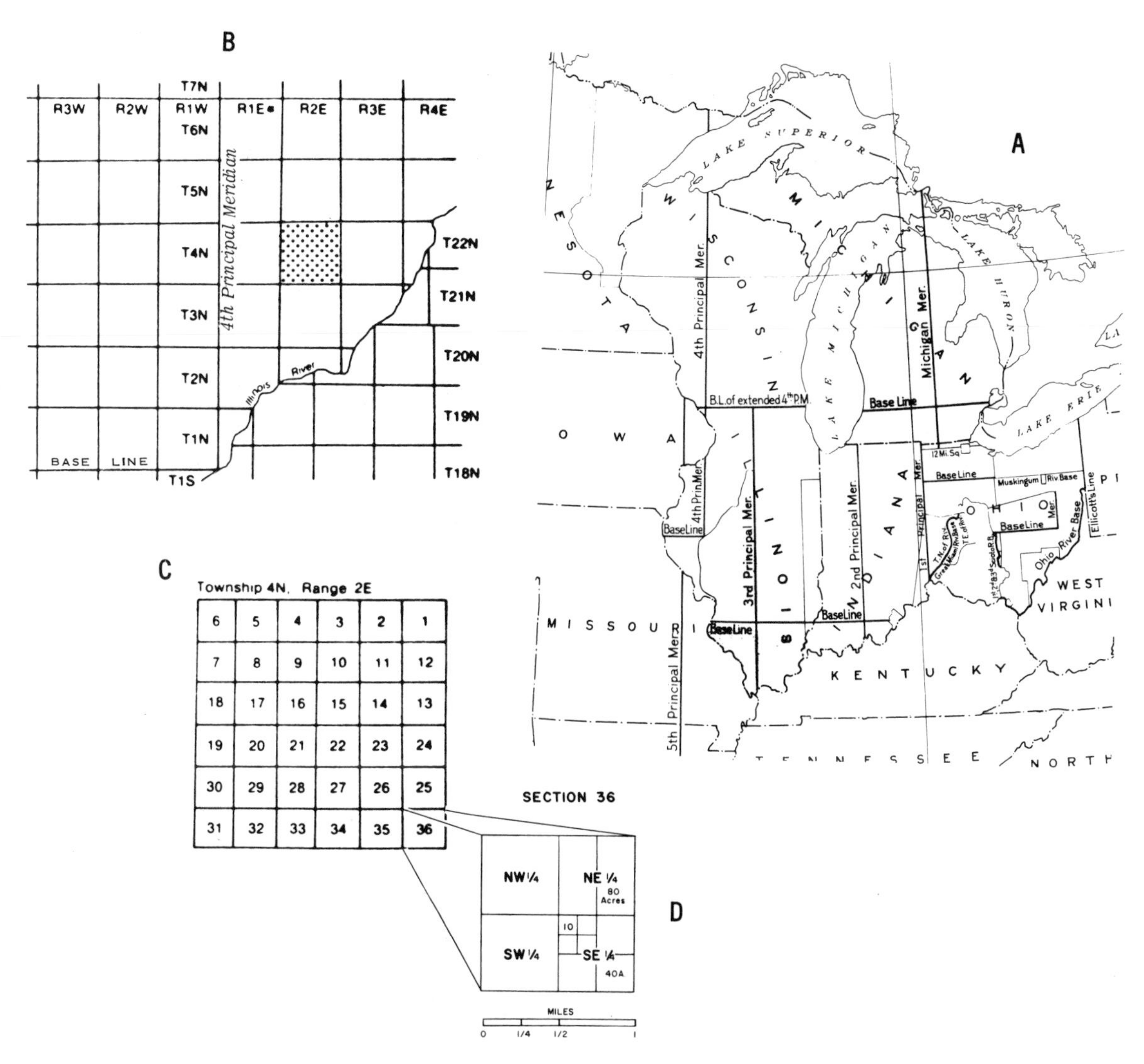

6	5	4	3	2	1
7	8	9	10	11	12
18	17	16	15	14	13
19	20	21	22	23	24
30	29	28	27	26	25
31	32	33	34	35	36

19. The Public Land Survey System: (A) principal meridians and base lines that form the framework for the rectangular grid system that comprises township and range land survey; (B) range and township pattern in western Illinois; (C) typical township that measures 6 mi. by 6 mi. on each side (note sequential numbering of the 36 sections); (D) breakdown of one section into quarters (160 acres), eighths (80 acres), sixteenths (40 acres), and sixty-fourths (10).

vary. Further, in fitting squares to the curved surface of the globe adjustments must be made, so some sections are oversize, some undersize. Interested persons may pursue the details in one of the references at the end of the chapter.

The location of a feature in field notes may be given within a ten-acre square by quartering the section in which it lies and then quartering it again and again. For example, the location of point 1 in part D of Figure *19* would be given as: C (center) of NE ¼ of the SE ¼ of the NW¼ Sec. 36 T. 4 N., R.2 E. When you go to find the point on a map you read the description backward, from right to left, that is, from larger to smaller area.

Latitude and longitude can be determined readily because the boundaries of these maps are determined by this global system. Each corner shows the correct geographic designation and tick marks subdivide the margins.

Highways, Byways, and Walkways

The scenic and natural beauties of Southern Illinois become more readily accessible year by year with construction of new highways and improvement of the secondary and back roads. The United States Forest Service, United States Department of Interior, State Department of Conservation, and other agencies supplement public roads with their own. Numerous walking trails are maintained, some marked by blaze symbols or even numbered locations keyed to trail guides. Appendix B lists the agencies responsible and a few outstanding examples of trails. Don't overlook the municipal parks and park programs.

Accessibility to a lovely area may lead to overuse and deterioration. The footsteps of hundreds of people trample the grass and tender plants. Slipping and sliding, especially when the ground is muddy, gouges out the roots and creates grooves for running water to erode. Whatever we do has an impact. Cutting or even collecting wood for a fire, throwing away a can or a piece of paper, digging one bloodroot plant out of hundreds—all leave their mark, multiplied a thousandfold.

In dead of winter I visited the site of a proposed nature preserve. The whole woods floor was covered by moss and lichen, even the path. I was enraptured by its beauty and wildness. We want to preserve it, but what will happen when public attention focuses on it? As nature lovers, we have responsibilities to be thoughtful of the places we love. We should strive to leave no sign that we passed that way so the beauty and fragile balance remain for the next visitor—and the next and next.

SUGGESTED READING

Tyner, Judith. 1973. *The World of Maps and Mapping: A Creative Learning Aid*. New York: McGraw-Hill.

United States Geological Survey. 1969. *Topographic Maps*. Washington, D.C.: Dept. of the Interior.

The Seasons

Summer

In summer grasses and crops mantle the open land, the leaves of trees obscure the cliffs and waterways, and broad vistas may be hidden by the growth. By midsummer most seeps and small springs and the low-order waterways are dry. In dry years even the big streams have little flow.

The hot dry periods are critical times for plant communities. This is the time to look for sites where the microclimate is critical: exposed cliff tops with bare rock and shallow soil; the west-facing hill prairies on silty loess; the alkaline "slick spots" of the flat prairie lands now cracked and brick-hard; the cooler, still moist north-facing bluff shelters. Now is the time to investigate caves and limestone springs whose temperature remains nearly the same summer and winter.

In our climate summer rainfall and evapotranspiration are indeed critical. Water stored in the soil and rock during winter and spring seeps out to the surface or is drawn upward and passes into the atmosphere. By summer the excess is gone. The ground becomes firm and dirt roads are passable. Seepage of groundwater stops. Eventually if there is no rain, plants begin to wilt because rootlets cannot draw enough moisture. When rains come the soil water is replenished; but there is seldom runoff except from scattered impervious surfaces and city paving.

The water level of ponds and lakes tends to decline in the summer, too. Minor wave-cut clifflets may form at the lower shoreline. In streams the more deeply scoured places still have pools of water where fish and water dwellers find refuge. In the channeled portions of the Saline and Kaskaskia rivers such water holes no longer exist though eventually natural processes will restore them.

Fall

The transition from summer to fall is gradual. Days may still be hot, the nights somewhat cooler. The average amount of rainfall is no more than in summer, but transpiration and evaporation rates decline. The ground is usually hard and dry. October is the month of low-stream flow. No water flows in the small waterways and even streams like the Kaskaskia and Saline rivers carry little water. In drought years the channels are dry except for deep pools.

The aspect of the land varies from year to year. The amount of moisture and the date of killing frost seem to affect the foliage color. In very dry years the leaves merely turn brown, whereas a rainy late summer and an early frost bring beautiful color.

Fall is not a time when active surface processes are readily apparent. The dry, cracked ground of summer takes up water and the cracks begin to close. Chemical alterations go on continuously but so slowly they go unnoticed. Rainfall is mostly absorbed by the soil, and runoff is nil or an immediate response to heavy rain. All but the largest springs are dry or a mere seepage. Exceptions to this are found in the Ozark province where the limestone bedrock reservoir yields water through the dry seasons.

When leaves begin to fall the ground is covered and rock exposures may be hard to see. Winds blow the leaves to sheltered places where they accumulate and may decompose to form humus. Surface runoff from heavy rains may wash much of the leaf cover downslope. The summer grasses die and turn straw-colored and tan. Blades and stems remain upright through the winter. Each clump of grass protects the soil surface beneath it. Cool-weather grasses turn green and together with fields of winter wheat provide a bright contrast to corn and soy bean fields and fall-plowed cultivated fields.

The alternation of warm- and cold-air masses dominate the weather pattern bringing warm, sunny days, and cloudy overcast or rainy days, and gorgeous clear, cold days. Gradually daily temperatures decline till fall becomes winter.

Winter and Early Spring

A walk through the countryside in winter brings particular pleasure to many outdoor-oriented people. The land is clothed in brown and gray. Trees and shrubs, bare of leaves, show their skeletal form and the color and texture of their bark. Rocks and cliffs can be seen through the woods and hilltop outlooks give better vistas. Pasture and unused fields sport patches of brown from broom sedge and little bluestem grasses. Tallgrass seed culms and seed pods of milkweed and other herbs grace the scene. Green of cedars, winter wheat fields, lichen, holly, all give an aspect to the land distinctive of these seasons.

Southern Illinois has a cyclical pattern of winter weather. Average temperature is close to freezing but cold temperatures are ameliorated by short periods of warm, sunny days when the thermometer reaches 60° F.—even in January. Typically we experience an alternation of moist, relatively warm air masses from the south followed by penetration of cold, dry arctic air from the north. Storms and rain or snow result but sparkling clear, cold days follow. The air masses drift eastward. Soon another wedge of warm air slides in over the cold air, bringing wispy cirrus clouds, an overcast, and then rain and low stratus clouds as the warmer air mass reaches us.

The active geologic and biologic processes affecting the landscape differ in winter from those in summer. Biologic activity and chemical changes are at low ebb. Evaporation rates are slow and removal of soil water by plant transpiration is slight. Rain is more frequent than snow. Unfrozen soils absorb the moisture so that the woods floor and pastures are wet though the footing is good unless overly impacted; cultivated fields tend to be muddy. Downward seepage saturates the ground and percolates downgrade until finally forced to the surface as seepage along the lower hillslopes and waterways and at rock exposures. Springs flow vigorously.

In colder weather the ground freezes. If the freeze occurs rapidly after a saturating rain look for frost heave, especially in the garden, cultivated field, overgrazed meadow—wherever soils are bare (Fig. *20*). Iris and young perennials in silty soils may be lifted right out of the ground. When a soil surface is covered by myriads of small pinnacles look closely for pillars of ice up to three inches high with a cap of soil, pebble, or twig. The ice is often banded. These ice crystals form after heavy rains have saturated the soil and the temperatures drop to twenty-eight degrees or below. Freezing water in the soil pores expands upward. As freezing extends deeper and deeper, additional freezing at the base of the columns forces the columns outward from the surface. If the next day is sunny and cold, sublimation occurs; the ice turns to vapor without melting and the pillars crumble. Commonly a peculiar lineated surface is left.

In cold periods groundwater seepages freeze and form terraces and flow features, icicles hang from cliffs and shelter bluffs, and waterfalls freeze over. In rocky stream beds look for slabs of rock wedged loose by the freezing of water which had seeped into joints and bedding planes.

Snow covers the ground nine to seventeen days per year on average. It brings special beauty to the landscape. The white mantle subdues the vista, but the delicate balancing of accumulated snowflakes on each limb and grass stem gives grace to each detail. Snow along streamways brings a fairyland, a mosaic of snow, water, ice, bare rock or ground, and bright green tufts of grass and moss. Best of all the snow remains white on the land. Few sources of dust or smoke exist to contaminate the pristine beauty.

Surface runoff is greatest during the spring, though in winter even small waterways usually have a flow. The saturated ground maintains the streamflow in the waterways even between rains. At times of heavy rain the water quickly flows away because soil pores are full. Bare cultivated fields suffer from sheetwash, rilling, and gullying. Large streams scour their banks and pick up coarse and fine sediment at places of high-energy impact and deposit it at places of energy dissipation. Floods are most likely from the overflow of small streams for short periods and from the backwater of major rivers, lasting for longer periods.

On hillslopes the ground may be unstable, especially where streams or man have oversteepened the base. Saturation reduces fric-

20. Ice pillars with cap of soil raised above the ground surface. Freezing temperatures following saturating rains cause ice crystals to form in the soil. Accretion of more ice at the base of the crystals causes them to grow upward above the surface. Note banding in fallen ice fragment in lower left.

tion between particles and increases the weight so as to cause slumps and mudslides and other types of mass-wasting.

SUGGESTED READING

Department of Commerce. *Climatological Data for the United States by Sections*. Monthly with annual summaries. Contains weather statistics from 47 separate sections. (Indicate desired section when ordering.) Asheville: National Climatic Center. One section $4.50 per year.

———. NOAA. 1973. *Monthly Normals of Temperature, Precipitation, and Heating and Cooling Degree Days 1941–70*. Climatography of the United States. Asheville: National Climatic Center. $0.25 each state.

Division of Industrial Planning and Development, State of Illinois. 1959. *Water Resources and Climate*. Atlas of Illinois Resources series. Springfield. 58 pp.

Flohn, Hermann. 1969. *Climate and Weather*. New York: McGraw-Hill. 253 pp.

Illinois Department of Business and Economic Development. 1967. *Water for Illinois, a Plan for Action*. Springfield. 425 pp.

Illinois State Water Survey. List of Publications. Urbana. Many reports on climate and weather.

Page, J. L. 1949. *Climate of Illinois*. Urbana: University of Illinois Agricultural Experiment Station, Bulletin no. 532. 364 pp.

Water

The River System

Willisville 1:24,000
Paducah 1:250,000
Alto Pass 1:62,500

How beautiful is the river system! The main stem gathering water from its tributaries, and each tributary in turn collecting from smaller and smaller watercourses. Viewed from the air a branching network covers the land. Yet most of the land surface is between channels. On this surface rainwater soaks into the soil or flows in a film from uplands and slopes. At the beginning of channels water capacity is small, but at each junction the channels become a little larger to accommodate the increasing flow.

Examine the map (Fig. *1*) and photograph (Fig. *21*). It becomes clear that rainwater joins a stream channel all along its course, not just at the headwaters. The source of a river at a single spring is a myth, true only for special cases in some limestone or volcanic terrains.

The river system belongs to humid regions, and to lands long exposed to rains and erosion. Arid lands may have rivers flowing through them. The Nile gathers its waters in the wet tropics, but crosses the desert in its own narrow valley without a tributary for hundreds of miles. Some arid lands have relict systems, integrated sand-filled valley floors mute testimony to wetter climates of the distant past. In mountains of a desert region enough rain falls to form streamways and valleys—but the valleys disappear in fans of sediment on the flat basin floor.

No river systems traverse newly formed lands. Rivers do cross deltas, emerging coastal margins, and drained lake floors, but the land is known more for ponds and swamps than tributary channels. Pleistocene glaciers covered northern Illinois 15,000 to 25,000 years ago. When the ice melted, irregular deposits buried the old valleys and drainage systems. Water filled the depressions and found its way between the ridges from pond to pond and finally into established valleys. Fifteen thousand years is not enough time for rivers to resculpture the land; only the main watercourses have eroded large valleys. In the Mt. Vernon Hill Country and Springfield Plain of Southern Illinois a more ancient ice sheet melted away, perhaps 200,000 years ago. Since then the processes of humid temperate climates have reestablished integrated river systems. In the counties along Interstate 70 broad flat prairie lands occupy the divides between rivers. Remnants of hilly glacial deposits and broad flatlands with hardly a water channel testify to the work of the ice sheet. Here patterns of surface runoff are discernible and headwater streams gradually extend their gullies into the land.

Each river system has many subsystems or drainage basins. Within each, surface runoff and groundwater seepage are collected and carried to the main stream. Some basins are symmetrical, others have tributaries mainly on one side. In some, the channels are rather widely separated, leaving broad uplands with a flat or rolling surface. Other basins have channels so close together that the land is mostly in slope.

The divides between drainage basins may be drawn on a topographic map between the

21. NASA ERTS (Earth Resources Technical Satellite) image showing the junction of the Mississippi-Illinois-Missouri valleys. Note the integrated river and stream systems in Illinois. The Kaskaskia River basin lies at the east, marked by Carlyle Reservoir. Smaller basins drain western Madison and Jersey counties. At top center of the image is Springfield Reservoir.

headwaters belonging to adjacent stream basins (Fig. *22*). Rainfall at the divide will run off in opposite directions. In the field the actual divide may be difficult to locate, especially if the upland is broad and flat.

The Mississippi-Ohio river system collects water from diverse regions, yet problems relating to flood or low water or seasonal flow require consideration of all their parts and their interrelationships. On a smaller scale it is still necessary to consider the characteristics of the whole drainage basin when planning a city reservoir or a conservation program.

Things to do. A river system or drainage basin is not readily identifiable in the field; therefore, use a map or photo composite or high altitude image. The parts of the system are best understood by going into the field. Waterways are physical features, but they are dynamic because of the energy of the water which flows in them.

Compare maps from different natural divisions. An interesting quantitative means of comparing drainage basins is to compute the drainage density. Measure the total length of all waterways (in miles) and divide by the number of square miles.

SUGGESTED READING

Illinois State Water Survey. Urbana. Many publications on rivers. Leopold, L. B.; Wolman, M. G.; and Miller, J. P. 1964. *Fluvial Processes in Geomorphology*, esp. pp. 131–50. San Francisco: W. H. Freeman.

Twain, Mark. 1883. *Life on the Mississippi*. New York: New American Library (1961).

United States Army Corps of Engineers. 1969. *Ohio River Basin Comprehensive Study*. 14 vols. Cincinnati, Ohio.

———. 1970. *Upper Mississippi River Comprehensive Basin Study*. 7 vols. St. Louis, Mo.

———. 1971. *Big Muddy Basin Coordinating Commission Comprehensive Study*. 7 vols. St. Louis, Mo.

———. 1971. *Wabash River Basin Comprehensive Study*. In cooperation with Wabash River Coordinating Committee. Washington, D.C. 112 pp.

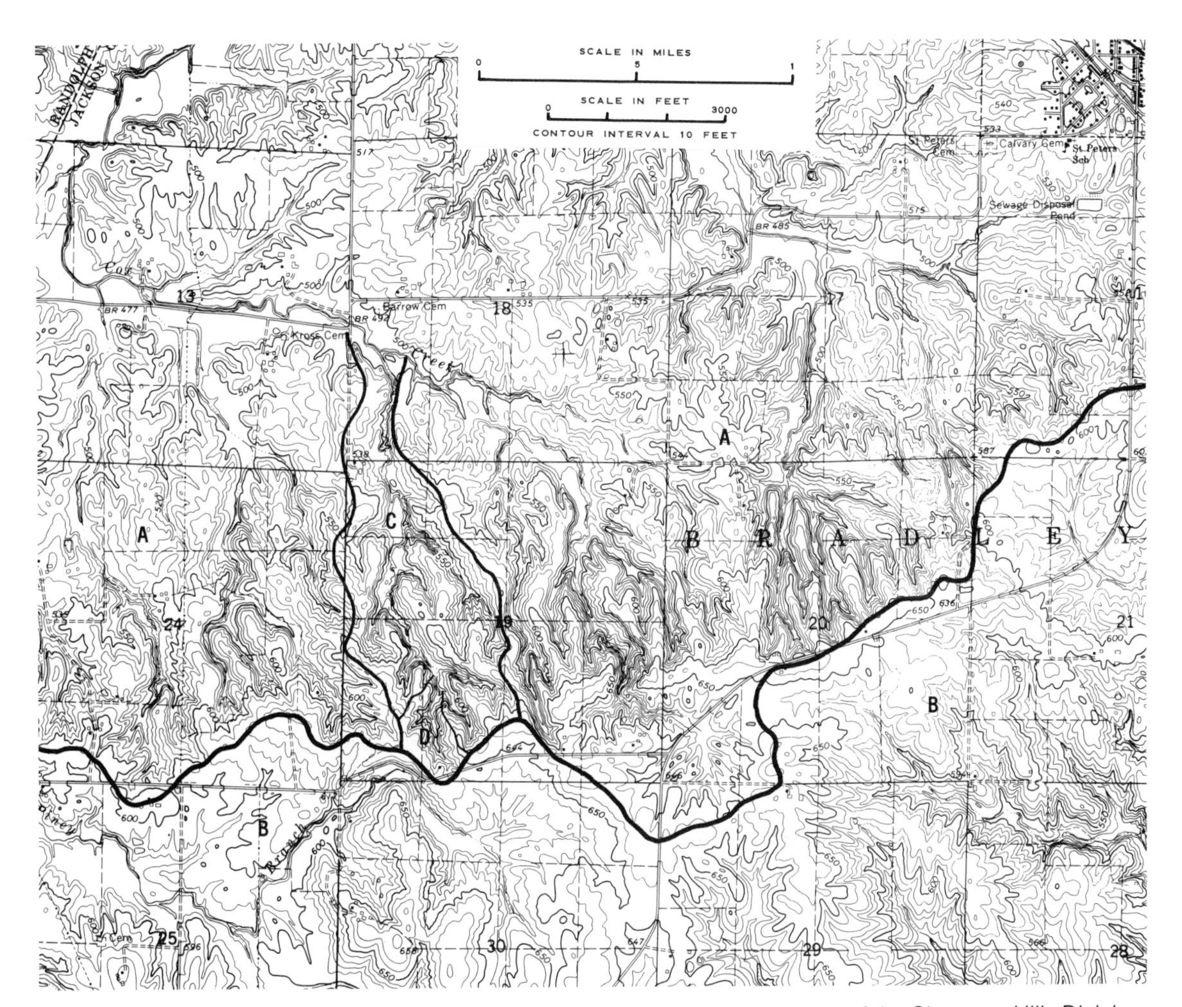

22. *Topographic map of the Willisville Quadrangle in the northwestern corner of the Shawnee Hills Division. The heavy lines show divides between drainage basins. Area D is a small basin with a third-order waterway and many small tributaries. Area C consists of several small basins and the master stream is fifth order. A major divide separates runoff between A and B.*

Overland Flow

Cisne 1:24,000
McLeansboro 1:62,500

The surface of the land is mostly smooth, else cultivation would be impossible. Flat, rolling, or steep, there is no immediate channelway to concentrate the rainwater which falls on it (see Figs. *23, 24, 25*). The first raindrops are held on the vegetation or are drawn into soil aggregates or clods. Some raindrops evaporate. Accumulating rainwater sinks into the soil pores. If infiltration is slow or rainfall intense, water puddles on the surface or slides downslope drawn to one obstruction and another. At last the entire surface may be covered by a thin film of water.

Go to a pasture or field during a rainy spell in winter or spring. The water-saturated soil or sod squishes beneath your feet. Get down close to the ground. Perhaps you can see the sheet of water as wide as the slope is broad. Movement is slow near the top of a hill where the slope is gentle and water accumulation is slight. Probably the water will be clear. Muddy it or put in some dye to trace its streaming laminar flow. No turbulence raises soil particles. Farther downslope more water gathers and velocity increases. Now some turbulence and water concentrating into channels becomes evident. The water may now be muddy. On a cultivated field there will be rills and gullies marking linear, concentrated flow. The farmer will smooth them over. On a good sod or on a matted woods floor erosion may be resisted and the water remains clear.

Raindrop impact and sheet erosion

On cultivated fields or other bare soils the impact of raindrops acts like a small explosion on soil particles. The splash picks up and disperses the particles. Billions of raindrops, each lifting a few particles. Accumulated raindrops make a film as they move downslope. The soil particles fall into the film making it muddy. In a single season tons of soil may be carried away by this process without leaving any telltale gully. This process is commonly called sheetwash.

Things to do. Drive a white board upright into the earth in a field. How high does the splash carry soil particles?

Devise a sediment trap to catch soil carried by the sheet of water flowing from a field. Compare erosion from a pasture and a cultivated field.

On flat uplands, channeled waterways may be miles apart. Measure the longest distance between road culverts.

SUGGESTED READING

Ellison, W. D. 1948. Erosion by Raindrop. *Scientific American* 149:40–45. (Offprint no. 817. San Francisco: W. H. Freeman.)

United States Soil Conservation Service. Many pamphlets available at local offices.

Seeps

Water seeps out of the ground along every stream channel and gully. This is most apparent in winter and spring when the ground is saturated. Such seepages are the source of continuing streamflow after surface runoff from rainwater ceases. Most seepages are very slow (it would take a long time to accumulate a cup of water) but they may extend along a valley slope for hundreds of yards (Fig. *26*).

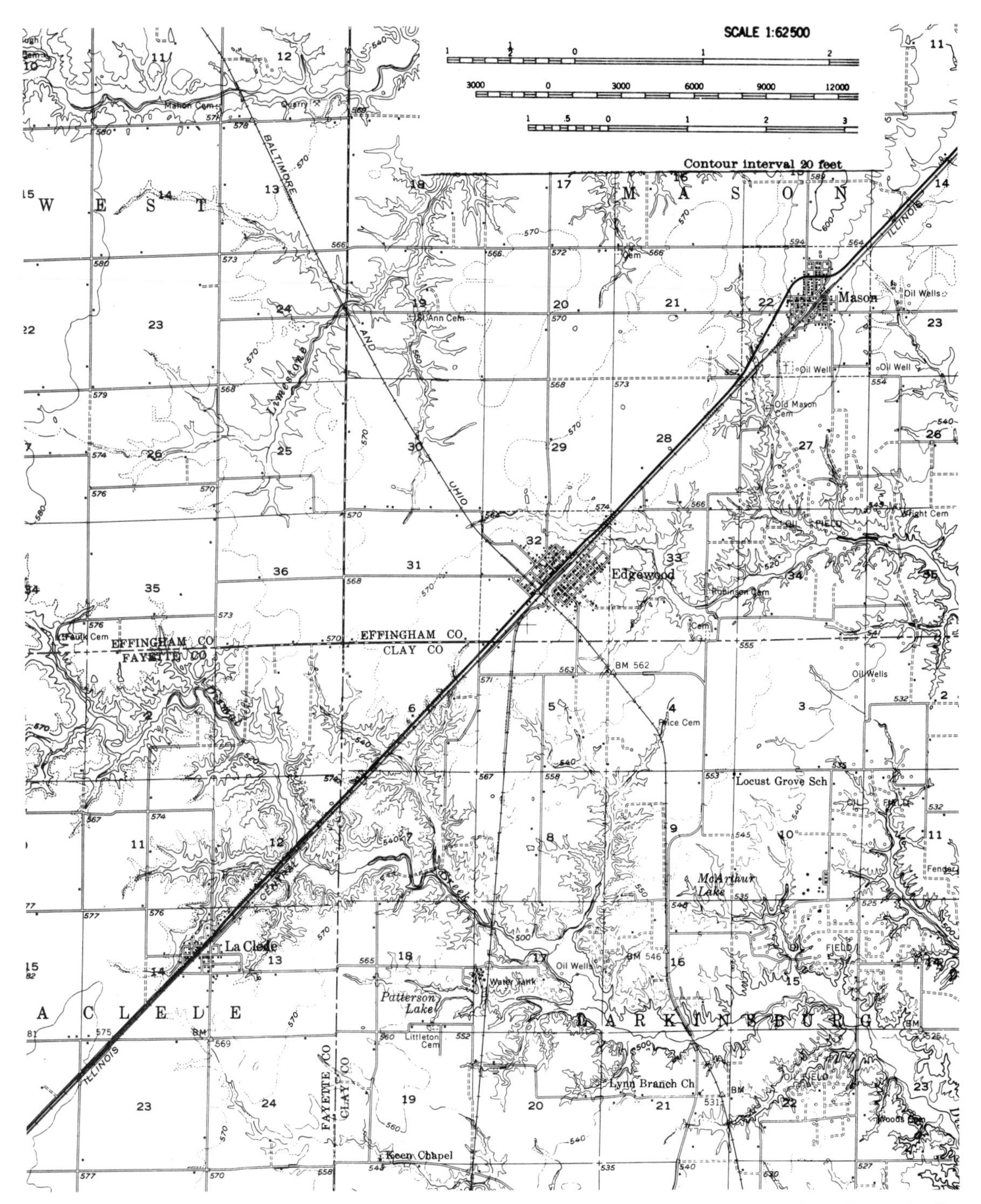

23. Part of Edgewood Quadrangle, Fayette-Marion-Clay counties. Highways I-57 and U.S. 37 follow the flat prairie upland. Divides between streams are broad, some without a waterway within a square mile. Closely spaced stream tributaries are eroding headward into the upland.

24. Looking eastward across gently rolling divide between Big Muddy and Skillet Fork Drainage basins in southeastern Jefferson County.

Rocky places yield the most obvious seeps. In Southern Illinois bedrock is mantled by windblown silty loess and glacial deposits. Rainwater seeps downward through this material to the nearly impermeable bedrock. Gravity continues to draw it downslope on top of the rock until it comes to the ground surface along the valley slope or to the stream channel itself.

It is pleasant to walk a rocky stream channel when snow is on the ground. Warm seepage water, protected by the earth's insulation, trickles out across the rocks and adds its bit to the stream (Fig. *27*). Bright green moss, sedges, and grass give their living beauty to the riverbank with its white snow and dark tree trunks. Gurgling water flows beneath ice patches and around and over angular rocks and boulders of the watercourse (Fig. *28*).

After leaves burst and vegetation draws groundwater for transpiration the seepages gradually diminish. By July or August few seepages can be found and the low-order waterways dry up. Groundwater still reaches the larger streams but mainly through the sands and sediment beneath the channel itself.

Things to do. Devise a way of taking the temperature of the seepage water. Will it have the same temperature as the soil through which it flows?

Try to collect water from a seepage area and estimate its yield in gallons per day.

Springs

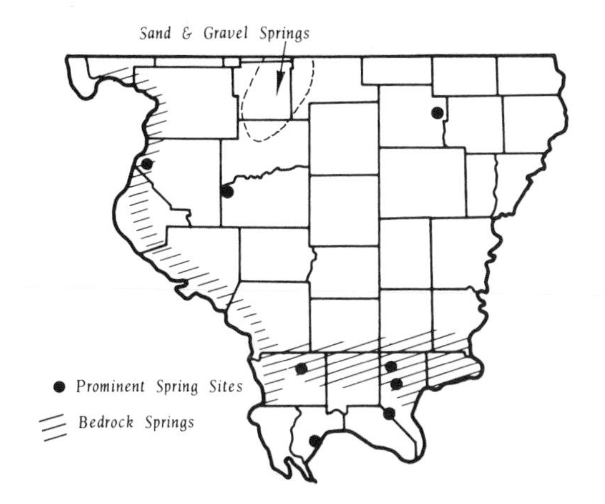

Springs are widely distributed, especially in the Ozark, Shawnee Hills, and Coastal Plain Divisions. Springs are less common in the plains area.

A number of large springs which flow the year round are well known in the limestone terrane close to the Mississippi Valley and in Pope and Hardin counties near the Ohio River. Some issue as small streams from caves while others rise along the base of the bluff or flow from small solution openings. In the Shawnee Hills many of the deep valleys have springs. They issue from the joints in the sandstone and also from fractured fault zones, as at Dixon Springs State Park.

The discharge from springs fluctuates with the seasons and with the cycles of wet or dry years. Most small springs go dry in the summer and early fall. During the drought of the early 1950s both streams and springs went dry. Yet the large spring at the base of the bluffs at Pine Hills, Union County, yielded enough water to maintain a large swamp area—thanks to the beaver which maintained a dam around it.

Springs in limestone terrane (Fig. *25*) respond quickly to rains. Surface runoff flows into sinkhole openings and into a joint and cavernous system below. The discharge at the spring is marked by muddy water. After the rain, water which has percolated through the soil also reaches the cave. Such water is clear, but may not be safe to drink. Where passages in the limestone are big enough to enter, one may observe the gathering of water seeping down the walls, dripping from ceiling cracks, entering along small side-passages. Springs in Shawnee Hills issue both from limestones and from the sandstones. Joints and fractures in the sandstone offer passageways for collection and flow.

Springs are most often found near the valley floor where the water table comes to the surface. An impervious shalelike layer on the hillside may inhibit downward percolation and cause a hillside spring. A good example may be seen beneath the cliff at the Alto Pass village park in Union County.

25. Spring issuing from a solution opening along the bedding plane of limestone by the side of Orchard Creek, Alexander County. People who live close-by still obtain drinking water from it.

26. Seepage of water along the top of a sandstone cliff on Fountain Bluff, Jackson County. Ground temperature was above freezing, but exposure to air temperature produced icicles and frozen waterfalls.

The Coastal Plain Division also has many springs. Some issue from a very specific orifice, perhaps from limestones which are exposed beneath the sands and clays. Others are described as seep springs. Water seeps out of the fine-grained white sands at a number of centers. Most are in steep first-order waterways. They seem to cause slumps of earth from the steep slope above. The fine sand is moved slowly outward but is held by ferns and sedges so as to make small terraces. The natural environmental conditions are stable enough so there is little erosion, but disturbance causes slides and instability.

Springwater is generally close to the average annual temperature, 56° to 59° F. A few degrees seasonal change may occur where water does not penetrate deeply or where it flows rapidly. Springwater has somewhat more dissolved solids than surface water. Most is calcium carbonate. Some springs were popular spas at one time. Dixon Springs, Saratoga, and Creal Springs all had hotels and attracted health seekers because of the mineralized water. Some springs along Lusk Creek also yield mineralized water. Springs are reported in some of the sandy hills of Bond and Clinton counties.

Things to do. When you find a spring examine the flow of water. Some comes up through sand causing little turbulent clouds

27. Seep of water dripping into a stream near Makanda, Jackson County. Water has infiltrated the soil and worked its way downslope at the soil-water contact.

28. Water flowing in January across a rocky bar near Makanda, Jackson County. Note imbricated (overlapping) sandstone blocks.

which rise and fall; some flow as a stream from a crevice. Perhaps you can measure the flow by collecting the water in a measured container, and note the time it takes to fill the container.

Longer term observations of fluctuations in temperature, cloudiness, discharge, and even mineral content would make interesting projects. Records relating rainfall patterns to discharge characteristics would make a valuable scientific study.

SUGGESTED READING

Ott, Christine E. 1966. Vegetation of Fresh-Water Springs in Southern Illinois. Master's thesis, Southern Illinois University.

Schwegman, John E. 1969. Vegetation of some seep springs in the Cretaceous Hills of Southern Illinois. Master's thesis, Southern Illinois University.

29. Gullies formed in loess and glacial till on a road cut. Gullies form rapidly on steep slopes where bare, loose materials are exposed to heavy rains, wetting-drying and freezing-thawing.

Gullies

Everyone tuned in to the outdoors and the land has a feeling of distress when coming upon a field being eroded by gullies. Many a cultivated upland field is scored by myriads of small gullies where soil has been scoured away by torrents of runoff after heavy rains. Deposits of eroded sediment may fan out at the mouth of the gully and bury the grass of a pasture below. Such gullies will probably be filled by the next cultivation and the process may be repeated until the topsoil is gone. Parallel gullies akin to these are seen on new road cuts or in areas cleared for construction (Fig. *29*).

The term "gullying" also applies to the trenching of waterways which had previously been a grassy or wooded swale (Fig. *30*). Such gullying is now characteristic throughout our region on the flat plains, in the bottomlands, and in the hilly regions. It indicates a "rejuvenation" of erosion adjusting to the new conditions caused mainly by man's activities. Removal of forest, breaking of sod, the straightening and deepening of main channelways, all increase the rate and volume of surface runoff. Removal of vegetation also exposes a more easily eroded material to the hydraulic action of the running water. Most gullying is in the unconsolidated mantling materials, the soil, loess, glacial deposits, alluvium. Rock may eventually be

30. Trenching of a swale southwest of Carbondale. Common throughout Illinois. Note "waterfall or knickpoint" at right. Sod forms a resistant lip, but turbulent water tumbles over the brink and undermines the sod. Retreat is very rapid.

exposed and trenching will actually cut into it.

Gullying does expose a cross section of the material eroded. Search out deeply eroded gullies to see profile sections of the soil, to look for the contact between glacial till and the loess, or to find out what the bedrock is like. Gullies may expose weak materials like shale and thin-bedded sandstones not usually found in natural exposures.

Discontinuous gullies

A very common feature of the uplands is the discontinuous gullies. They generally occupy the swale of a first- or second-order drainageway. Such gullies characteristically have a low waterfall a foot or more in height at their head, a steep-sided trench for some yards, followed by an untrenched fan deposit. A second gully may appear a short distance downslope. The falls and gully side are held up by resistant sod. The sod is undermined by scouring of the slits beneath the densely packed roots until the sod collapses, causing the gully to retreat headward. The eroded silts are soon redeposited, filling the lower reaches of the gully.

Things to do. Gullies are natural laboratories for the study of erosion by running water and even mass-wasting. They represent an unstable and dynamic condition where change is taking place. Miniature river basins can be mapped, photographed, and watched during a period of time. Bulldozed construction sites offer excellent models. Waterfalls, headward erosion, meandering or braiding of stream channels, depositional features, pirating of tributaries, and so forth, can all be watched and the processes studied. Such an area can be a valuable asset for a class in geology or earth science or for an individual study project.

Stream Order

Willisville 1:24,000
Wolf Lake 1:24,000
Iuka 1:24,000

At some critical distance and gradient of slope, the accumulation of overland sheet flow is concentrated into turbulent linear flow; here waterways begin. The map (Fig. *22*) shows a series of parallel waterways all heading at the same relative position on the slope. Each small waterway leads into a similar channel and into successively larger streams. The orderly arrangement and speacing of these waterways is striking when once noticed.

From the upland or ridge-crest, walk downslope across a field or pasture. Soon a little collecting basin appears where sheet flow becomes concentrated into a linear path. This is the head of the drainage—the first-order waterway. Look along the slope and you may

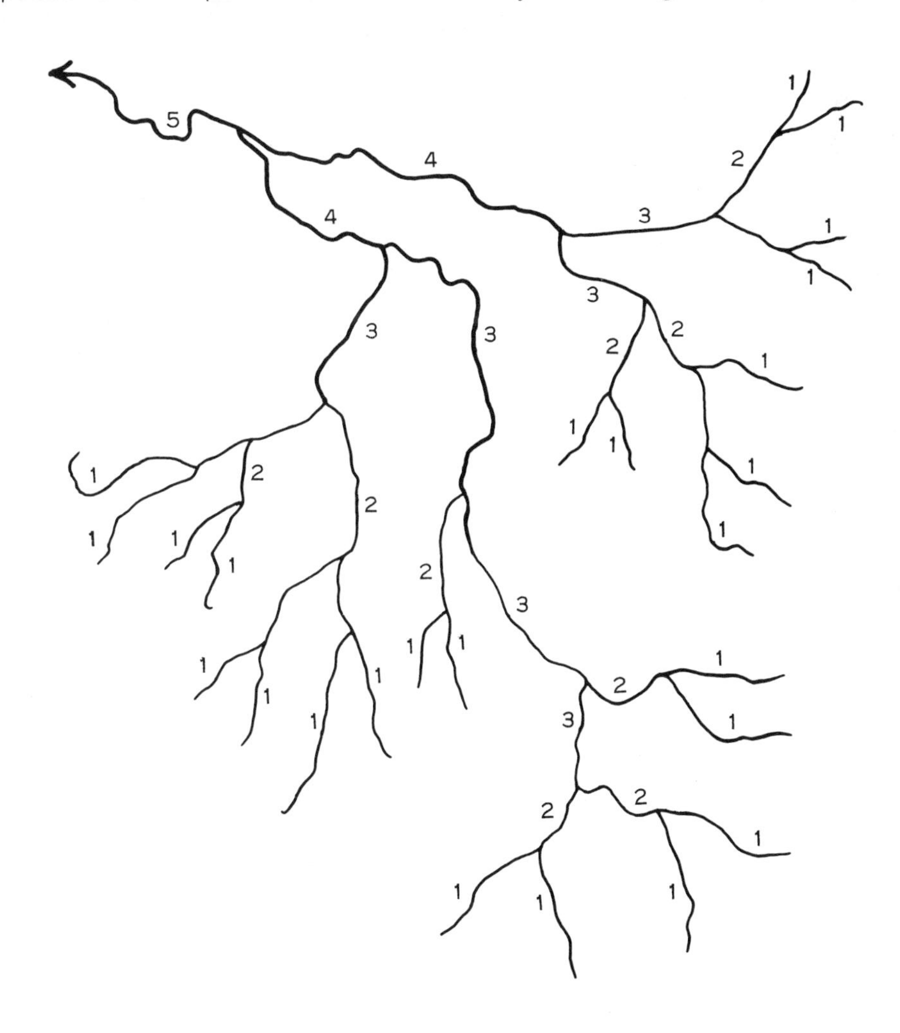

31. Stream Order.

32. Cultivated field with two first-order waterways joining to form a second-order waterway tributary to a higher-order main channel. These low-order waterways are trenched, but waterways in untilled area are merely swales. Note small gullies and also lighter colored soil where sheet erosion is active.

see several such collecting areas, rather evenly spaced and one much like the next (Fig. *31*).

The very first waterway at the beginning of a concentrated flow is designated a first-order waterway. It has a concave collecting area causing the overland flow to concentrate along the axis. It 1) has no tributary; 2) carries water immediately after rain and then goes dry; 3) may be grassed or wooded without even an eroded channel bed. Two first-order waterways join to form a second-order waterway (see Fig. *32*). The latter must be a little larger to accommodate water from its two sources, and a definite channel is common. The junction of two second-order channels increases the order to third, and two third-order channels produce one of the fourth order. Lower-order tributaries may enter at various places but order magnitude would not be increased. The diagram (Fig. *31*) and map (Fig. *22*) illustrate these designations.

Channel, or stream order, may be applied to any region. It is often quite a chore to identify the order of the largest rivers, but it is quickly done for low-order streams. The concept is useful because where soil and rock, relief and slope are similar, the stream features in each order are similar to one another. In the field notice the width and depth of the channel, the sediment or rock on the floor or lining the sides, also its gradient or slope. Size and spacing of

33. Looking upslope along a low-order drainageway. The first-order collecting basins can be seen at each side of the barn.

waterways are related to the volume of water which must be carried by the average annual flood. In turn, the amount of runoff is regulated by the intensity of storms on the one hand and on the slope and and infiltration capacity of the soil on the other. In winter and spring the pores of the soil are commonly saturated with water or perhaps filled with ice. Rainfall or snowmelt at such times must all run off. Transpiration from plants is slight and evaporation minimal in cold weather. First-order waterways achieve the capacity to carry this runoff while second-order channels must have the capacity to transmit the flow of at least two first-order waterways (Fig. 33).

Things to do. Low-order streams offer much of interest. They are often dry and always small enough to cross easily. Look for seeps and springs, the first exposure of bedrock or glacial deposits. Here, too, are most waterfalls and enlarging or deepening gullies. Comparison of waterways and valley slopes of different orders may be interesting. Note also the differences in the vegetation cover and man's use of the land.

SUGGESTED READING

Strahler, A. N. 1965. *The Earth Sciences*. 2d ed. (1971). New York: Harper & Row. 824 pp.

Intermittent Stream Channels

Lick Creek 1:24,000
Fairman 1:24,000

Low-order water courses are dry much of the year (Fig. *34*). Most of the time one can walk along the channel. Stream channels of each natural division are basically similar yet differ from the channels of other provinces. Channels are usually trenchlike features along the bottom of a valley. In areas of steep slopes, low-order channels may occupy the entire valley floor but through most of Southern Illinois channel banks are bordered by a flattish floodplain.

Natural channels are sinuous, turning from one side of the valley floor to the other. Most channelways are dynamic, perhaps because of fluctuating water volume. When the channel is dry, mud shrinks and cracks and animals burrow. In floodtime the loosened material is wrenched away and even large blocks may be tumbled or scooted along. Some channels receive much eroded sediment, from clay to boulder size. During high waters these are carried downstream, only to be deposited as velocity decreases. Deposits are found overbank, on the inside of the bends, in bars of silt or sand or gravel. Obstructions such as a log or growing willows cause deposits against or behind them. A fallen tree may force the current to one side causing undercutting of the bank and a change of channel position (Fig. *35*).

Stream channels exhibit alternate stretches of scour and deposit or pools and riffles. Along one stretch the channel may have a smooth floor perhaps scoured to bedrock (Fig. *36*). Sand, pebbles, even boulders are swept across this surface finally to be dropped in a bar or against one another or obstructing trees. Deep pools are comparable to the smooth channel floor. Most of the time the water is still and quiet, but at floodtime the current rushes along carrying all its load of sediment right on through; look for fresh deposits downstream from the pool.

Notice the kinds of sediment in the stream channel. They have fallen or slid into it or were washed in by rain and tributaries. These are the materials which are available. The different kinds of sediment and pebbles found in the stream bed are the ones that will be found exposed upstream.

34. Intermittent stream channel in Shawnee Hills, Union County. The waterway is dry in late summer. Note bare rock reach of the channel and holes where sandstone blocks were ripped out. In foreground is a bar of deposited sandstone blocks. These were probably transported in the previous year.

35. The stream channel is dynamic. Here the bank is being undermined and scoured on the outside of a bend. Overlapping blocks show the current moved from right to left.

Look at the pebbles. Most will be the resistant rocks of the area, sandstone in the Shawnee Hills, chert (flint) in the Ozarks. In an area which may have been glaciated, look for erratics (wanderers) of granite or gneiss or "greenstone." If you find platy fragments of easily broken shale there must be an eroding exposure a little way upstream (Fig. *37*). Pebbles of light-colored limestone may only indicate that crushed rock is being eroded from a roadway.

Relatively few rocks will be found in the stream channels of the Plains Division. Fine materials line the channel sides, which means that bedrock is not exposed. Streams in hilly lands also carry much fine material, but you may have to look for it in quieter places or as a thin coating after the water has receded. In such streams the fines do not accumulate except, perhaps, to fill in the open places between the gravels.

Where there are slablike blocks, notice the way they have been tilted up against obstructions or against each other on a bar deposit. The imbricated arrangement as shown in Figures *35, 36, 37* and the diagram (Fig. *38*) tells the direction of the current flow. You may be able to find such blocks in the channel bank, representing a deposit of long ago.

In a rocky stream channel blocks which are in the process of being loosened can be found. Note that the rock is layered into beds (Fig. *39*) and that vertical cracks (joints) are common.

36. Dry stream channel in September in Giant City State Park, typical of the rocky Shawnee Hills Division. Bare rock is clean of pebbles and boulders; all have been scooted across the smooth surface. They pile up in bars. The power of storm discharge can be imagined from the slabs of rocks piled against each other.

37. Stream flowing against the valley wall and eroding a weak shale. Shale blocks are found for a short distance downstream, then disintegrating pebbles, and finally only little shale flakes. The nature of the pebbles indicates the kinds of rocks and sediment that lie upstream.

38. Imbricated Boulders.

Plant roots, burrowing animals, freezing and thawing of water all tend to widen the cracks until finally during floodtime the block is pushed out or torn loose by the deep and turbulent stream.

Rapids

Turbulent waters of rocky rapids bring beauty and interest to small streams in the hilly areas where bedrock comes to the surface (Fig. *40*). To enjoy the rapids one must walk in winter and spring because rapids are found only along the intermittent portions of streams. Rarely does water flow the year round except in very wet summers. The Mississippi River is an exception. At low water go to Chain-of-Rocks,

39. Thin-bedded sandstone in creek bed, Giant City State Park, Jackson County. Note how beds wedge out and overlap one another.

40. Rapids in a Shawnee Hills stream near Makanda, Jackson County. The thin-bedded sandstone is eroding rapidly also. The large blocks have come downslope from cliff at left. The boulders in stream channel have been caught there as high waters receded. They will soon be moved again during a storm.

Grand Tower, or Thebes Gap where the great river has a rocky channel. Rocks, abraded and smooth, line the shore and protrude far out into the channel. Occasionally barges cannot pass through the shallow rock-filled water. At high water the confined river increases in velocity and the rocky riverbed can only be guessed at by the boils and standing waves.

Rapids are formed where the stream crosses bedrock more resistant than that immediately upstream or downstream. Below a rapids the valley floor widens out and the gradient of the stream is more gentle. Rock is probably no longer visible. Above the rapids there may also be a wider valley and a quiet reach of the stream.

Rock appears in the stream bed at the head of rapids, generally a massive sandstone, but in the Ozark Division the bedrock is limestone. Each rapids has its own character, and the flow of water takes different forms with varied discharge volumes. The gradient is relatively steep and the velocity of the flow rapid and turbulent. Some rapids consist of a series of small steps, but rapids are distinguished from a waterfall by a continuous slope rather than a free fall of water.

Look for details. Smooth surfaces face upstream abraded by speeding sand and pebbles, a sharp, beveled edge lies above the protected down-facing side. On the latter, surface moss and lichens are protected from the friction and impact of transported sand and cobbles. Joint and fracture patterns of the bedrock are clearly visible, and offer access to water and plant roots. Look for blocks or slabs being wedged apart or perhaps for places from which a block was recently loosened.

Potholes or moulins are present in many stream beds, notably Orchard Creek in Alexander County. The intersection of two joints have been abraded and eddy currents cause sand and small pebbles to circle round and round. Centuries of abrasion gradually enlarge and deepen the hole in solid rock. Most moulins have sand and pebbles in them; these are the tools. They grind and are ground down, and eventually become so small they are carried away. A limestone moulin will abrade rapidly; a sandstone moulin more slowly. The grinding pebbles in a limestone moulin are most probably of flint and very hard, whereas limestone is quite soft by comparison. The tools in a sandstone region are probably also sandstone pebbles, perhaps well cemented by iron oxide or silica. Granite and quartzite pebbles from the glacial till are found in some moulins, especially in the western prong of the Shawnee Hills in western Jackson and Randolph counties.

Rapids often occupy a confined segment of the valley. At floodtime water becomes surprisingly deep and swift. Twigs, branches, and other debris caught in trees and shrubs record the water level. You may wonder at the dead leaves and twigs, even branches, caught in the trees head high—a caution that the channel itself does not have the capacity to transmit maximum runoff.

A stretch of rapids is nearly free of sandbars, pebbles, and boulders, even though the channel above and below is well filled with them. The rather regular hard-rock surface and steep gradient make an efficient conduit. The fast-moving, turbulent water pushes along every particle and boulder which reaches it. Just like a playground slide, nothing stops along the way.

Immediately below a rapids there is often a pool scoured in the stream bed. Beyond the pool the debris comes to a halt in a large bar of cobbles and boulders of all sizes (Fig. *36*).

Waterfalls

The waterfalls of Southern Illinois are not spectacular but the setting of most is wooded hill lands off the beaten path. The best known are in parks or, like Burden Falls (Fig. *41*), Pope County, close to an all-weather road. Thus the hiker in hill country often has the pleasure of making his own discovery. Waterfalls are found throughout the Shawnee Hills and Ozark divisions and in the scattered rocky hills of the Mt. Vernon Hill Country. All are on intermittent

41. Burden Falls, Pope County, in summer. Only a trickle of water streams over the ledges. The lip is massive sandstone, but below it a weak shaly zone erodes more rapidly.

42. Small waterfall across the Tar Springs sandstone in Union County. The force of falling water excavates the shale below the sandstone and scours out all the debris carried over the lip.

streams so that a fall of water is found only in the wet months or during rainy periods.

Stream channels ordinarily have regular slopes varying somewhat in steepness of gradient. A sharp interruption of the gradient in the form of rapids, cataract, or falls adds interest and beauty to the stream although transport may be impeded. To the float or canoe enthusiast they all bring excitement.

A true waterfall has a vertical drop with accompanying plunge pool below excavated by the force of the falling water (Fig. *42*). The same hydraulic force carries debris out of the pool where it is deposited as a rocky rim on the downstream side. Cataracts consist of a series of small vertical drops differing from the chute or rapids which have steep and rocky slopes but no true vertical falls. All these are often associated and one may evolve into another.

The resistant ledge at the lip of the falls is essential to its origin and its persistence. The resistant rock often extends along the valley wall forming bold cliffs or canyon walls. It also inhibits scouring of the stream channel. Look for weak rock beneath the falls. Clay shale beds or platy siltstones have many fractures and bedding planes which bring about disintegration by wetting and drying, and freezing and thawing. The hydraulic force of a stream in flood easily erodes the resulting fragments. As the weak rock is carried away the overlaying resistant cap is undermined and blocks of the lip collapse. Very large collapsed blocks remain in the plunge pool until they are reduced

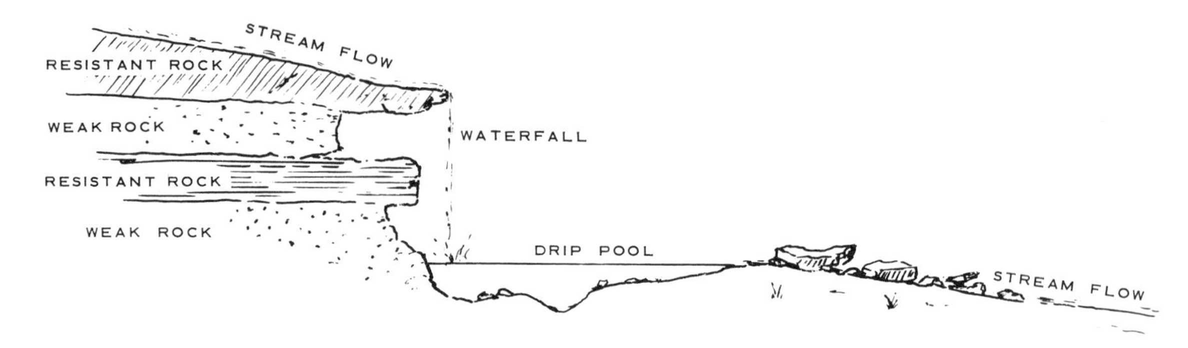

43. Waterfall Profile. The lip of the falls is a resistant rock, but this will soon break off because the underlying weak rock has been eroded so far behind it. The plunge pool is also eroded into weak rock. The turbulence of the falling water in flood clears the pool of the rocks that tumble into it.

in size by impact from other blocks falling upon them and the scour of sand and pebbles swirled against them; little by little fragments are washed away until the collapsed block disappears from the pool.

A waterfall remains just so long as a plunge pool in weaker rock remains below the cap. Eventually the fall retreats upstream until the structural relationship between the rock layers and stream gradient no longer allows removal of debris from the pool. Blocks, then, are not undermined and a vertical fall is not maintained. A transition to a chute and rapids begins. The head of Cove Hollow in Jackson County is an example.

The diagram (Fig. *43*) shows the resistant-rock unit extending under the hill. Weak layers cause reentrants in the face of the falls, and the hydrologic power of the falling water has scoured out a plunge pool at the base. Collapse of the resistant layers causes upstream retreat of the falls.

Waterfalls are the product of dynamic conditions. The land must be hilly and composed of materials of differing resistance to erosion. Furthermore, the processes of valley deepening must dominate over depositional processes. Some falls are the direct result of some event of the geologically recent past. Niagara Falls came into being because glaciation so greatly altered the Great Lakes region that the runoff is ponded in a series of deeply ice-scoured basins. Overflow from one basin to another occurs at the lowest sag. Between Lake Erie and Lake Ontario the sag led across the Niagaran escarpment.

In Southern Illinois most falls have a less dramatic origin. They probably stem from a time of the deepening of the Mississippi-Ohio-Tennessee river valleys, perhaps two hundred feet below the present floodplain, near the beginning of the Pleistocene glacial epoch. Previous to that the river valleys had more gentle gradients. Deepening of the main stems caused all the tributaries to become deeper and wider. These processes encountered the resistant massive sandstones of the Shawnee Hills. Cliffs and canyons were formed along the valley sides and waterfalls on the watercourse.

Retreat of the main tributaries left their smaller tributaries hanging also. Some of these have produced a considerable reentrant into the valley wall, but the lower-order tributaries merely drop across the cliff in their path into the canyon below.

A similar phenomenon is traceable to man's activities. Deforestation and cultivation cause more rapid runoff and channel deepening. More water reaches the waterways quickly during heavy rains, in turn requiring larger and/or steeper channels. Deepening of the main watercourse produced a headcut or knickpoint along all the tributary waterways. These are found throughout Southern Illinois. The unconsolidated sediments, loess, glacial till, Coastal Plain sands and clays are especially vulnerable.

The practice of deepening and straightening stream and river channels to promote more rapid runoff is bringing about more profound changes. Only recently has the seriousness been admitted by the conservation and waterway agencies who are no longer recommending new projects, yet those already approved will be carried out. Deepening of the main stem leaves all tributaries, including road ditches, hanging. The steeper gradient produces a retreating knickpoint. Straightening the channel reduces its length and hence steepens its gradient. At times of high water, velocity increases and the capacity to erode causes greater deepening and widening of the man-made trench. A visit to the Kaskaskia or Saline valleys, recently "improved," provides instruction in river dynamics as the flow ranges from a mere trickle in summer to a flood in spring.

Straightening of the Cache Creek channel in Pulaski County and shortening of its length by means of the Post Creek cutoff is making some profound changes in the Cache lowland, as shown in Figure *44*. Indeed the original swampiness has been reduced, so cultivation is possible. However, dissection of the bottomlands is accelerating and Heron Pond Nature Preserve is endangered as the Cache Creek channel has deepened to fifteen feet below pond level.

Things to do. When you visit a waterfall examine the rock. Note the differences in weathering and erosion of the layers. Take pictures from marked locations and return in later years to observe whether erosion has been active.

Go to a falls during or right after a heavy rain. (In many places bluff shelters beside the falls will keep you dry). The increase in flow creates a spectacular sight as the stream of water shoots outward to the middle of the plunge pool.

Meanders

The curving and twisting channels called meanders are characteristic of streams on floodplains (Fig. *45*). Streams on very young flat surfaces also meander. During low flow the stream curves back and forth within its own channel. Strangely, the meandering pattern appears to be the most efficient means of handling discharges which vary greatly from season to season, or even week to week. At maximum flow the channel cannot carry all the water, so it overflows the bank onto the floodplain. This overflow water takes a direct route between bends. As the water recedes some is retained in low places which serve as temporary storage areas.

The curving channel has a much lesser gradient than the direct route, so water flows more slowly but efficiently withall. Artificially straightened channels soon experience undercutting and sliding of the banks as obstructions cause the current to move from side to side. Note how much wider a straightened channel is than its meandering forebear. Note also that shorter distance causes steeper gradient, faster current, and ever deeper scour. Visit the Cache River at Post Creek cutoff or the "endangered" swamp at Heron Pond Nature Preserve in Pulaski-Johnson counties.

SUGGESTED READING

Leopold, L. B., and Langbein, W. B. 1966. River Meanders. *Scientific American* 214:60–70. (Offprint no. 869. San Francisco: W. H. Freeman.)

Floodplains

Most streamways, fourth order or higher, have a floodplain (Fig. *45*). Typically in the Till Plain Division the channel is trenching itself deeper into the alluvial fill and in the Shawnee Hills and Ozark divisions has encountered

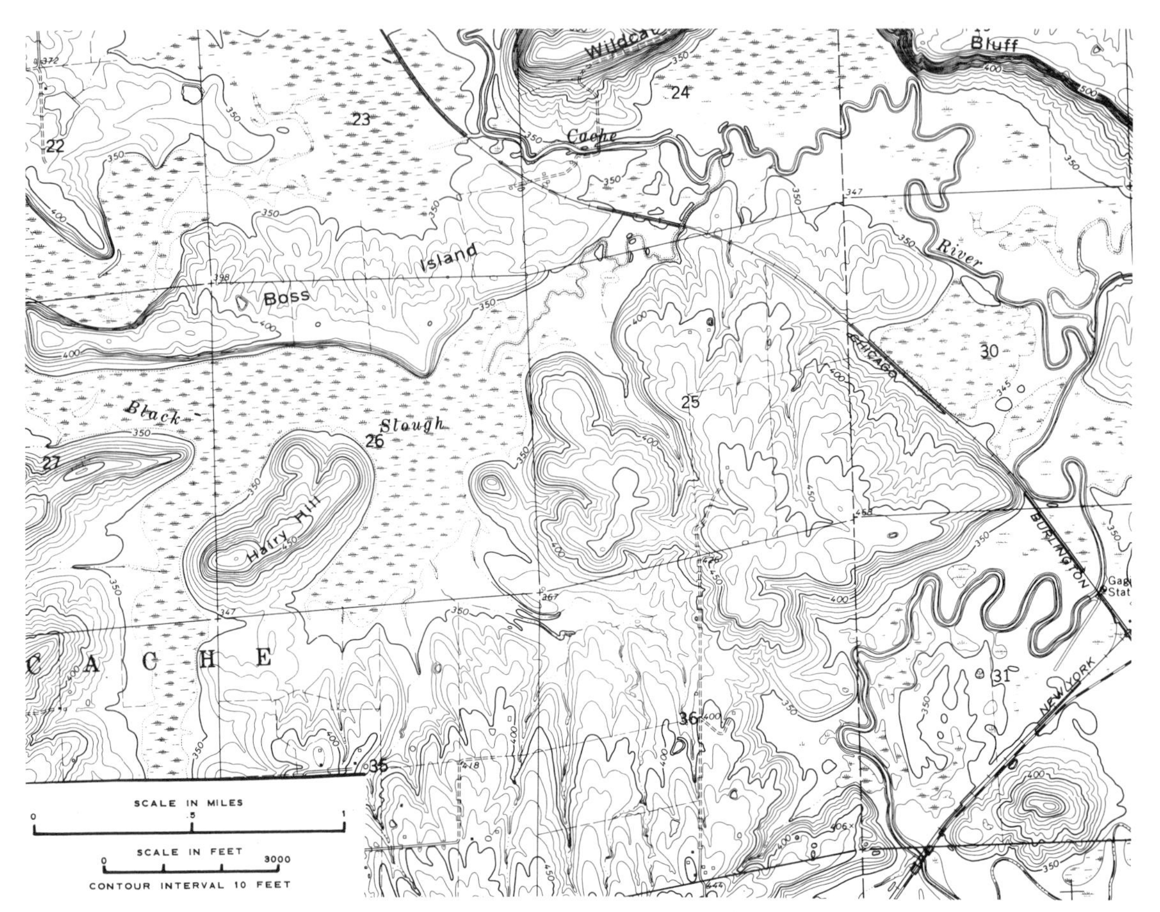

44. Part of the Karnak Quadrangle, Johnson County, showing Black Slough and Cache Creek between high hills at the north side of the abandoned valley once occupied by the Ohio River.

rock. The floodplain is really not very flat for it contains overflow scour channels and raised bars of sand, silt, or cobbles. Secondary channels at the edge of the floodplain carry along the waters from the valley sides, perhaps for a long way before reaching the main channel.

Most floodplains have several distinct levels. An older one at a relatively high level may rarely be flooded, and successively lower surfaces contain more frequently scoured channels. These seem to be the result of rather rapid deepening and shifting of channels, some since the settlement of our region.

Take a walk along the floodplains of a small stream in the spring. Wild flowers cover the woods floor and water flows in the channels. This is the time of greatest contrast between the levels. The highest ones have the deepest accumulation of litter and the lushest cover of flowers. Lower levels have suffered more recent erosion, so there may be only a scant cover of vegetation.

Where the channel flows against the valley side the floodplain is widened by undercutting (Fig. *46*). Slide features are often seen. Such places expose the rock or sediment of the valley slope. This is the place to look for weak bedrock which may not ordinarily be exposed.

45. Floodplain with sun glinting off flowing and standing water. The trees that line the twisting stream meanders mark the present channel. An abandoned, partially filled meander channel retains water in the cultivated fields from middle to right edge of photograph. The farmer has ditched fields to hasten runoff, but he has not straightened the stream channel. The edge of the floodplain is clearly marked on the valley slope where gullying terminates. Washington County.

In the plains divisions glacial till and boulders of crystalline rock are exposed.

Stream Sediments

Sediments are carried along a stream channel in high water but are deposited as velocity decreases. It is interesting to plot the distribution of sediments and their textures within the stream channel. The channel is deepest on the outside of the bends and along narrow straight reaches. On the inside of the bends is a sloping deposit called the point bar. Gravel and cobbles are often found highest on the bar and the finer material below.

Another bar is located at "the crossing" near the middle of the straight reach between bends. Mark Twain describes the nature of the river channel and its sediments in a colorful way in *Life on the Mississippi*. River pilots follow the deep water around the outside of a bend but must cross a shallow bar to the deep water on the opposite side at the next bend.

At floodtime the stream frequently overtops its channel banks, especially across meander bends. Sediment is carried out of the channel and deposited on the floodplain. The coarse material is deposited near the bank, building a natural levee. Sand and silt is deposited across

46. Stream channel near Hurst, Williamson County. No rock is exposed. Note the slump along the channel side. The channel crosses a narrow floodplain in the right middle ground.

the floodplain. Some of the sediment may be washed down directly from the upland.

Alluvial Fans

At the foot of gullies and small tributaries deposits of the eroded material are laid down. The sudden decrease of stream gradient at the main valley causes a loss of velocity with resulting deposition. The deposition causes water to be diverted and the sediment is spread out in a fanlike shape. Fans are topographically higher than the rest of the valley bottom. Some are high enough to be above flood level and large enough to be favorable for cultivation. The texture of material reflects the nature of material carried from the upland. Fans are most common where there are steep slopes and considerable relief.

Stream and River Channels

Mt. Carmel 1:62,500
Okawville 1:24,000
Thebes 1:24,000

Channels of larger streams differ from intermittent stream channels only in degree. The flow of water follows the same annual cycle of high waters in late winter and spring and low waters in late summer and fall. The large bounding rivers never go dry, but during drought all the others are reduced to a trickle or a series of pools. During such times there may even be reverse flow up the tributaries from the master rivers.

The channel of Kinkaid Creek (Fig. *47*) below Kinkaid Lake is typical. It is an efficient U-shaped conduit with smooth silt-lined walls. Rarely can coarse-grained deposits be found within the channel. The surface materials of the region are fine grained and rock is rarely exposed. Near the Shawnee Hills, cobbles and boulders are brought to the stream channels by tributaries, but they are quickly transported through the conduit. When viewed at low water

47. Kinkaid Creek near junction with Big Muddy, Jackson County. Channel is narrow and U-shaped, lined with silts; no gravel bars. It is fairly stable, but slump on left bank middle ground has carried small tree into the channel.

48. Riffle across a gravel bar in an Ozark stream, Hutchins Creek, Union County. The gravel is chert (flint). Ozark-type streams flow year around, except during severe drought, because the limestone bedrock acts as a sponge and reservoir.

it may be hard to imagine that there could be enough runoff to fill the channel, yet overflow onto the adjacent floodplain is almost an annual event.

Hutchins Creek (Fig. *48*) is a beautiful example of streams of the Ozark Division. The channel is relatively broad and shallow. An abundance of chert pebbles is brought to the stream by each tributary. This debris forms bars on the inside of every bend and crossing bars between bends. Deep pools are scoured on the outside of bends and against limestone bluffs; channel banks expose both fine and coarse debris.

There is much to see along a stream channel. Look for places where the bank has slid into the channel or the roots of some great tree resist the turbulent action of the water. See where a flood current has carried sand and gravel over the bank, or where quiet backwater has left a muddy veneer on the bottomland.

The larger streams all have floodplains (Fig. *49*) across which the channel meanders. The channels are dynamic. Bank erosion, especially on the outside of the meanders, is continual. The eroded material is transported downstream a short distance and then deposited. A great load of sediment is transported. Coarse material is moved mostly as bed load in the main channel. Dredges remove it for use as commercial sand and gravel. After high water and floods sand is left on the banks. The sus-

49. Meander of the Wabash River, White-Gallatin counties, showing successive scrolls and swales resulting from shifts in the river channel. As the channel erodes the outer side of a bend, deposition takes place on the inside, thus maintaining the same channel capacity.

pended load is silt and clay. Tremendous quantities are carried away each year—the erosion products of our cultivated fields and bulldozed construction sites.

At low water the sediments left on the banks display current features such as ripple marks, channel scour, eddy scour behind obstructions, and drying features in the mud such as polygonal cracks and mud curls. These reveal alternate layering of fine- and coarse-grained material.

Things to do. The United States Army Corps of Engineers publishes river channel navigation charts. These are kept up to date annually and successive editions show repeated changes. The Corps attempts to stabilize the channel with levees, bank protection, and groins (Fig. *50*). Maps may be obtained from the Crops or seen at the map libraries at Southern Illinois University and University of Illinois.

A visit to a major river is most rewarding when one looks for sediment features and observes the flow of water. Visit Ft. Defiance State Park at the junction of the Ohio and Mississippi rivers. Watch the waters of the two rivers come together—and flow side by side with swirling eddies, but remain unmixed. The rate of flow of the Mississippi might be measured in low and

50. Groins of piling and rock extend out from the shore of the Mississippi River near Chester, Randolph County. The groins tend to restrict the channel width and help maintain a deep channel for barge navigation.

flood flow. At flood it carries an astonishing amount of floating debris including great logs and tree trunks.

SUGGESTED READING

Leopold, L. B.; Wolman, M. G.; and Miller, J. P. 1964. *Fluvial Processes in Geómorphology*. San Francisco: W. H. Freeman. 522 pp.

Valleys

Mt. Carmel 1:62,500
Kimmswick 1:62,500
Vienna 1:24,000

Valleys have been made by the streams which flow through them. One wonders how a small intermittent stream flowing only in its narrow channel can excavate so deep a feature. It may be hard to believe that the excavation has been accomplished by the same processes

51. Upper tributaries of a valley in Perry County. Note the way tributaries and gullies join at a common level.

active today; but the stream itself is dynamic and there is time in the establishment of the major valleys—thousands of years, perhaps millions of years.

The small stream we usually see does little work, but at time of flood it scours its bed and banks; it may erode against the valley wall itself. Many small tributaries and gullies carry rock and soil from the valley side into the main stream channel. Slides and slumps and sheetwash bring material downslope. Once this debris reaches the channel, floodwaters carry it out of the valley.

Valleys differ from division to division. Mississippi, Ohio, and Wabash bottomland divisions are themselves great valleys. The valleys of the Shawnee Hills are typically rock walled, narrow, rocky and steep in the headwaters, wide downstream. Ozark valleys are similar but with less rock exposed. In the Central Lowlands the valleys are more open and shallow with gently sloping sides.

The photograph (Fig. *51*) shows the upper course of a valley. Everywhere the slope is toward the stream. Each tributary seems nicely adjusted and joins the main stream accordantly. Of course, it could not erode deeper than the master. Hard layers or recent deepening might cause a drop from tributary to master—but rarely. The channel is small, yet a flat bottomland has already developed. Downstream the channel increases in capacity to accommodate the waters brought by many tributaries. The valley itself becomes deeper and wider. The valley slope is not directly affected by the stream below it. Yet every fragment carried to the bottom of the slope by gravity, sheetwash, and gullying is eventually removed by the stream.

Steepness of the valley slope depends on the relief of the area and the resistance of the rock which forms it. Where rock is resistant it may form steep cliffs, but where it is weak and subject to rapid weathering and erosion the slopes are commonly very gentle. Pounds Hollow in Gallatin County has high bluffs, whereas the valley slopes of Big Muddy or the Middle Fork of the Saline River are very gentle.

The floors of some valleys are very narrow with room for little more than the stream channel. Some flat places are usually present and as the stream gets larger there is generally a floodplain. The name itself implies that the plain is merely the highwater overflow channel. The landowner or municipality which fails to heed this warning eventually gains firsthand knowledge. Most large valleys have terraces which do not flood. Terraces are remnants of the former valley floor. Deepening by the river has eroded much of the old floor leaving flat-topped remnants (see Figs. *65, 66, 67*).

Bluffs

Valmeyer 1:24,000
Wolf Lake 1:24,000
Chester 1:62,500

Three hundred miles of high bluffs give outlook across the major river valleys. Stand on a high bluff or atop a spectacular cliff and view the valley spread out two hundred to four hundred feet below (Fig. *52*): the river scouring the base of the cliff or merely a tantalizing glint across the plain; flat and rich farmland laced by intersecting curves of abandoned meander bends, some wet and wooded, some deep, water-filled oxbow lakes; discontinuities in land use reflecting relief, drainage, and soil types; levees with their white-surfaced gravel roads paralleling the main channel, yet curving back to the bluff along each tributary. Straight lines are highways and railways—the cars and trains appear mere toys from that vantage point. Smoke plumes mark power plants and industry. Near the cities man's structures, stores of materials, and piles of waste cover the plain. Yet see how cultural use reflects nature's pattern.

Such outlooks can be found at hundreds of places in the land shown on the index map. Where no path leads to the cliff top one might imagine himself to be the first to stand there, or that last before him an Indian contemplated the river and its wet bottomland forest.

Look down. The cliff falls away and the tops of trees come into view. Outward and upward a buzzard or hawk soars against the open sky. Look along the bluff (Fig. *53*). In profile the eye is drawn to the bare rocky cliff with scattered plants clinging to ledges and crannies and to trees whose roots grow back into cracks and bedding planes. A steep slope of deep buff-colored silt clothed with shrubs, trees, and grass rises above the cliff. Down below is a steep slope of debris, fallen or washed from above.

Each bluff site is unique yet these three elements of landscape—cliff, rounded crest, debris slope—can be recognized (Fig. *54*). One may stand directly on bare rock above the cliff

52. Mississippi Valley from Pine Hills, Union County. The river is on the far side of the valley. The road is on the top of the levee raised to protect the floodplain from backwater up the Big Muddy.

or, where the cliff is modest or even absent, the best lookout is from a steep grassy hill prairie gracing the bluff top. Near Belleville and Edwardsville the Pennsylvanian bedrock is not resistant, so the bluffs consist of wash and debris slopes capped by thick windblown silt (loess); little rock is exposed.

When visiting a river valley bluff search out a good view from below and from a distance (Fig. 55). See how the parts fit the topographic map (Fig. 56) so that a closeup examination by road or trail will fit into place. A frontal assault is rarely practical because debris slopes are steep and unstable. The cliffs, especially limestone cliffs, are so fractured that hand and footholds easily break away (Fig. 57). In the spring, blocks have been loosened by the repeated freeze-thaw of winter. An overhanging cliff, quarry, or road cut is especially dangerous when the sun warms up a frozen rocky slope. The best access to the bluff top is from a tributary valley and the side or back of the cliff.

Examined from below the prominent horizontal layers segmented by vertical fractures (joints) are clearly visible (see Fig. 57). The rock is sedimentary, either sandstone or limestone. Thick layers tend to weather more rapidly as water, air, plant roots, and animals penetrate the fractures and wedge the blocks apart. A reentrant forms in the cliffs where weak layers have eroded away; eventually the overlying massive blocks, lacking support, tumble to the debris slope below. Vegetation is sparse on the rocky cliffs, but certain layers transmit

53. Profile view of the Mississippi bluffs at Pine Hills, Union County. Compare with diagram, Figure 54.

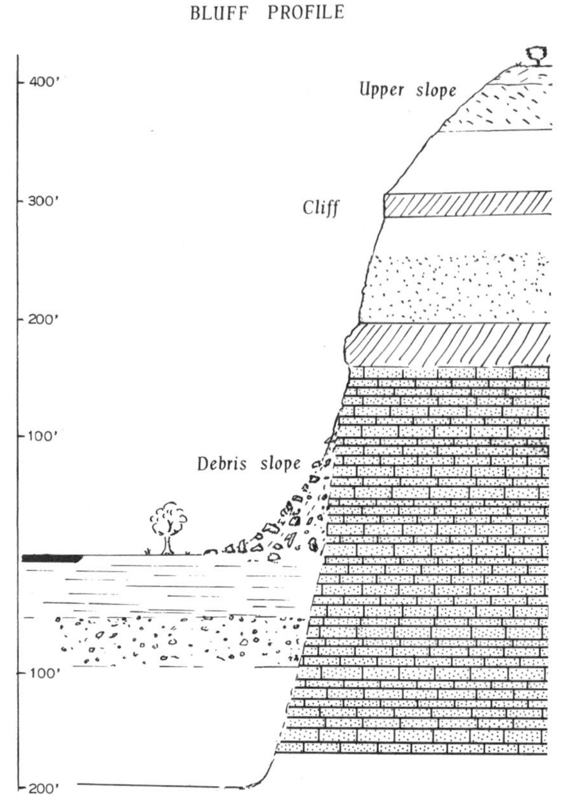

54. Bluff Profile. River bluffs consist of three parts. The nearly vertical rocky cliff is most prominent. Above the bare rock a steep slope of loess covering weaker rocks extends upward to the crest. Beneath the cliff fallen debris makes a talus slope that becomes more gentle outward to the floodplain. The large valleys have a thick fill of alluvial sediment above the bedrock floor.

water or retain moisture causing bands of grass and herbs. Small trees, often beautiful in configuration, cling to the face, their roots seeking moisture and nutrients in fine materials sifted into the joints and bedding planes. Look for seeps of groundwater, marked by algae and moss, and springs and small caves. Between the limestone mines at Prairie du Rocher a spring gushes water thirty feet above the road and a larger spring discharges at road level. This spring is the discharge of a long underground stream. The nearby mine broke into the cave whose waters at times of heavy rains give much trouble.

The cliff is not always readily accessible from below, because steep debris slopes mantle the base of the bluff. The slope consists of blocks and fragments tumbled from the rocky cliff, and sand and silt washed off the crest. This is a dynamic surface, with last year's fallen blocks precariously balanced against a tree or another block. The slope is at the "angle of repose" such as a pile of bricks dumped from a truck. Most debris slopes are wooded. Debris slopes beneath limestone cliffs are more active and perhaps more dangerous than those be-

55. The valley wall southeast of Fults, Monroe County. Compare with diagram, Figure 54. *A wooded debris slope rises steeply from the floodplain. The rocky cliff ranges from 50 to 100 feet high. Above the cliff is a steep hill prairie rising to the wooded crest.*

neath sandstone cliffs. In some places the debris has been partially removed in road building or for fill material. An example of such instability is attested by repeated damage to Illinois 3 south of Chester.

One wonders at the great size of some blocks (Fig. *40*) and whether they fell and rolled to their present place of repose or whether they gradually slid downhill. Some boulders disintegrate in place as weathering breaks them apart. The smaller fragments are more readily transported.

The upper part of the bluff is less prominent yet its steep slope may rise to twice the elevation of the cliff top. Such high slopes consist of rocks too weak to form cliffs—shales and impure or thin-bedded sandstone and limestone. They are everywhere mantled by loess. The loess is especially thick on the Mississippi bluffs although bedrock may be exposed in the short, steep tributary valleys. Glacial till lies between bedrock and loess in the sections where glaciers reached to the bluffs.

Bluff tops are generally wooded; some natural hill prairies exist on very dry western exposures. The prairies are used for grazing and the wooded slopes for timbering. Some houses have magnificent views in the few places where the general upland extends to the valley side.

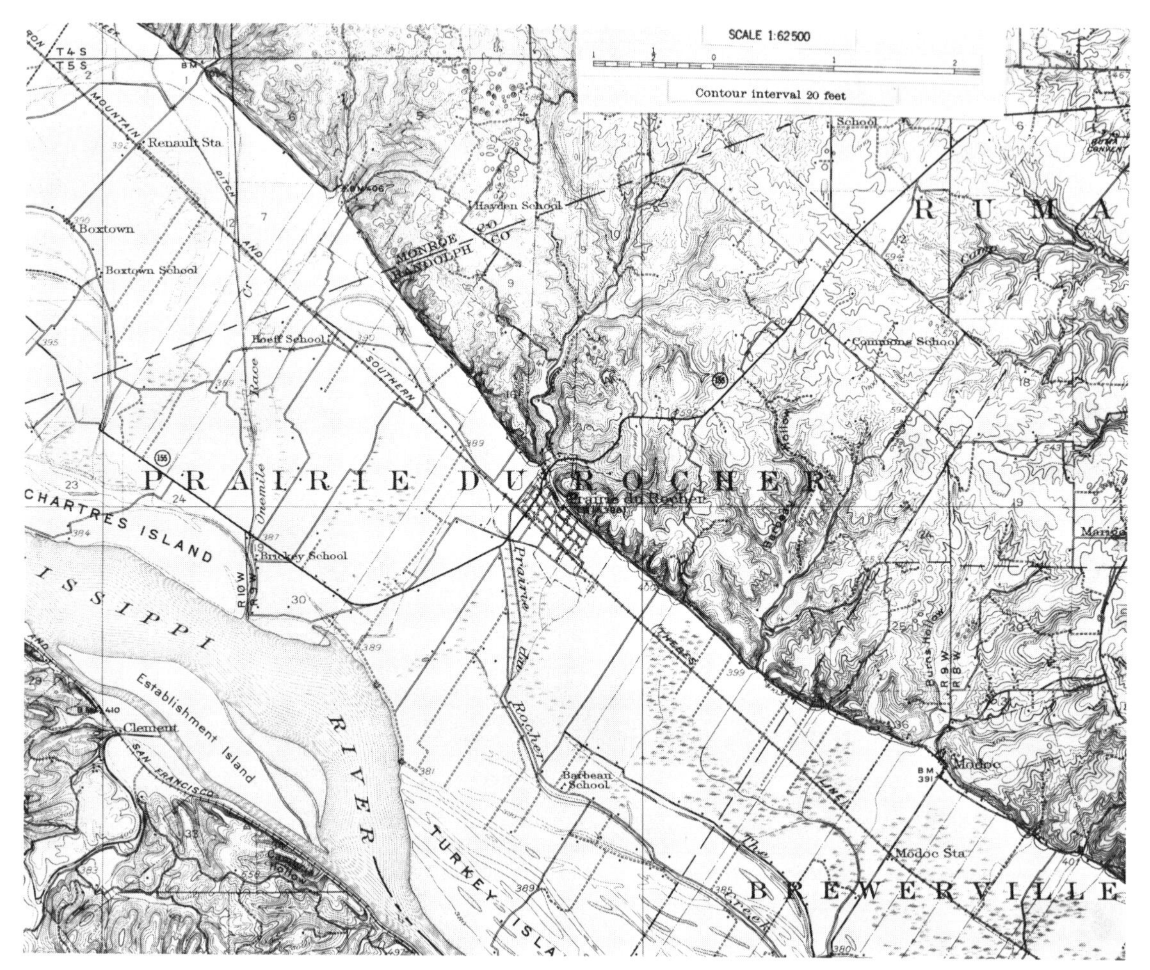

56. *The Renault Quadrangle illustrates the steep bluff and stream-dissected margin of the Mississippi Valley. The upland is part of the Ozarks and is dotted with sinkholes. See photographs, Figures* 112 *and* 113. *Floodplain swamps and curved waterways identify abandoned meander bends.*

Floodplain

Wood River 1:24,000
Cache 1:24,000
Shawneetown 1:62,500

Someday when the rivers are in flood make an airplane flight across Southern Illinois. It reveals more than a map, which shows only where floodplains are and their most obvious features (Fig. *56*). The rivers are full, and bounded by narrow levees, natural and manmade. Wide muddy backwater parallels the rivers and floods the tributary junctions; curving ponds are everywhere making scrolled patterns separated by similarly curving ridges (Fig. *58*). Most farm structures occupy the high places, and highway and railway embankments keep traffic flowing (Fig. *59*). Even the towns may be flooded in the severest floods, as in the spring of 1973 along the Mississippi River (Fig. *60*).

At low water the aspect is different but the forms are the same. The rivers are shrunken within their channels, swamps or marshes occupy the backwater areas, though deeper oxbow lakes persist, and most flood-filled ponds are nothing but dry swales.

A floodplain is the product of the river (Fig. *61*). The river's heavy load of sediment—sand and mud washed from the land upstream—is carried over the bank during high water. Once out of the turbulent channel the velocity decreases and semiments drop out of the floodplain. Especially in quiet backwater places new mud buries the old surface and raises the level of the plain. Abandoned channels are gradually filled and backwater swamps eventually are raised above the water table. Just as the river brings sediment so it scours and takes it away. Most erosion takes place on the outside of the river bends where the current cuts away at the high bank. Scour is from below and the

57. Looking upward at the cliffs of Bailey Limestone at the Pine Hills, Union County. The cliff is an active environment down which storm waters flow, against which summer sun shines mercilessly, and from which rock fragments are pried by winter's freeze and thaw. The steep talus slope of fallen debris has been partially removed for road fill.

58. Floodwaters of the Mississippi River, 1973. The natural levee next to the river in left foreground is well above the flood. Segments of old meander bends are marked by the curving bodies of floodwaters.

unsupported bank caves in . This process often affects the highest surface of the floodplain. The original village of Kaskaskia occupied such a site. Piece by piece it was, long ago, eroded away.

Floodplains are extensive in our area. The Mississippi, Ohio, and Wabash rivers are notorious for their floods, but we can add the riverless Cache-Bay creek lowland and all their tributaries. Careful surveys of Franklin County classify one third of all the land as subject to flood. Here the "bottomlands" extend along the Big Muddy River and even along the small tributaries.

A natural river course winds (meanders) back and forth through its floodplain although technological man seems to think it should be straightened (Fig. *62*). An undisturbed channel is difficult to find in Illinois. Nevertheless, even where channels have been ditched the tendency to meander can be seen and the processes are at work that cause it to do so again. The same processes are as active in a small stream as in a large one.

The meandering stream pattern is the key to the landforms of the floodplain. Find a small unspoiled meandering stream preferably unwooded (Fig. *63*). On the outside of every bend is a steep bank where the floodplain sediments are exposed and evidence of recent undercutting can be seen. Maybe a slump block partially obstructs the channel, and close observa-

59. Kaskaskia Island at the height of the 1973 flood of the Mississippi River. These are higher areas.

tion will reveal curving cracks on the ground above—presaging another slide. On the inside of the bend no cut bank exists. Instead, the ground slopes very gradually upward to the higher surface of the floodplain. Signs of deposition in the form of small ridges of mud, or sand and sticks and branches, line the banks. Indeed much of the material eroded from the outside of one bend is deposited in the quieter water of the inside of the next bend downstream. This deposit is known as a point bar. Point bars build into the channel as fast as the bank is eroded on the outside, keeping channel capacity constant. The curves and swales visible at floodtime or on an aerial photograph reflect the shifting channel and the growth of point bars.

A little search of the stream will reveal a place where high water overflowed the bank and followed a path across the inside of the meander. That pathway might also receive drainage from the valley sides. If the floodplain is broad so that the river meanders do not reach from bluff to bluff, the floodplain drainage does not find its way across the relatively high levee which immediately parallels the channel. Drainage must parallel the main channel until a meander comes against the valley wall. A striking example is the course of the Big Muddy. It enters the Mississippi floodplain opposite Gorham but hugs the east valley wall in a twisting course for twelve miles. There it suddenly turns across the floodplain and enters the Mississippi River. The map and aerial

60. Kaskaskia Island, Randolph County, at the height of the Mississippi River flood in the spring of 1973. The portion of the levee in the foreground is amply high and well riprapped with large limestone fragments. Floodwaters entered through a break at a weak place.

photograph (Fig. *64*) show why. Formerly a big Mississippi meander curved eastward from Grand Tower right to the base of Pine Hills and then back again to the Missouri side. Big Muddy, which previously had continued its course through LaRue swamp and down to Clear Creek, was intersected by the big river bend.

The very processes which keep extending the meandering curve of the channel eventually cause abandonment of the wide bend. Overflow across the neck of the bend causes scouring of the shorter channel. At first the river flow goes both ways, but water moves more slowly around the longer bend causing sediment to drop out and bars to build across the old channel. The river cuts off the old bend which is left as an oxbow lake.

Occupance

Why does anyone make his home on the floodplain of a great river? The name itself is a promise; it will flood. The farmer dwells there because of level land and productive soils. The urban dweller was attracted to the river for transport and water, to the level land, then to the highways and railways connecting major cities. At first dwellings and businesses were mostly on the higher ground of terraces and natural levees. The age of technology seemed to promise control by means of raised levees, drainage ditches, and flood control dams. People felt it was safe to occupy the lower areas, and increasing population pressure accelerated the move despite the fact that flood insurance was too costly to carry. Severe floods in the 1960s and early '70s inflicted great damage and hardship when levees failed or when they retained so much internal water that the land flooded anyway. Federally backed flood insurance is now available but the guidelines call for reduced construction in lowlands which are subject to floods.

SUGGESTED READING

Matthes, G. H. 1951. Paradoxes of the Mississippi. *Scientific American* 184:18–23. (Offprint no. 836. San Francisco: W. H. Freeman.)

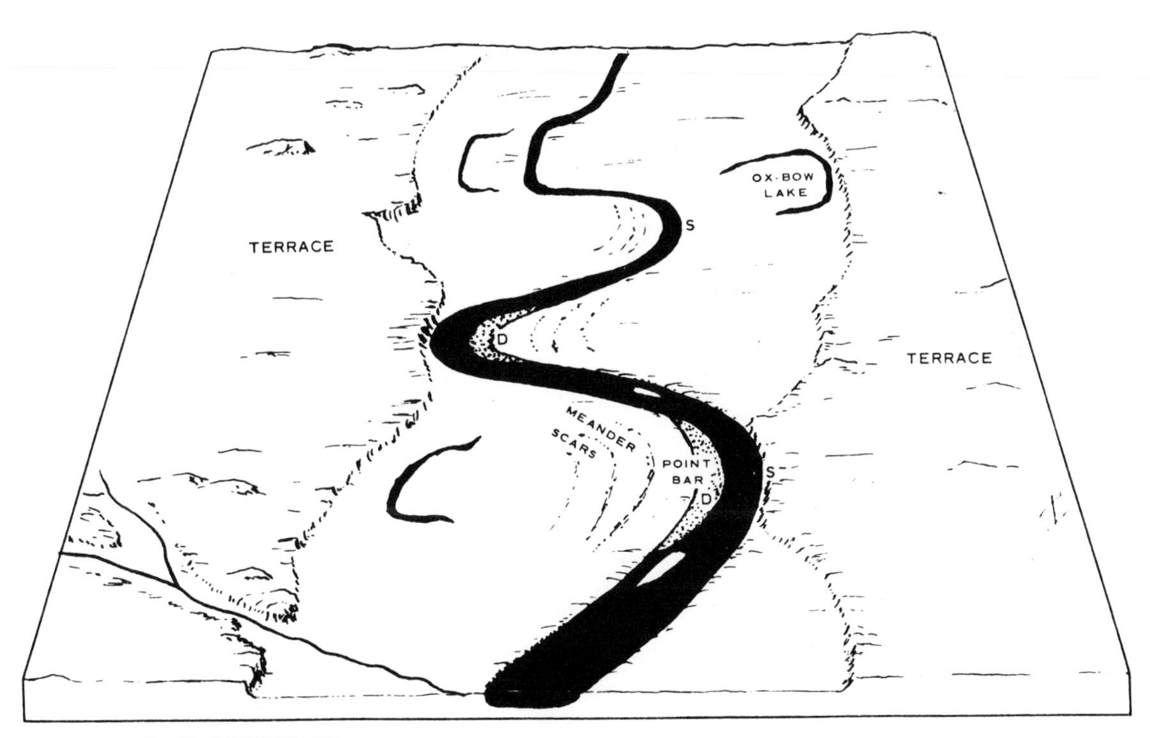

61. Floodplain Features: floodplain and meandering river channel shows the locus of erosion (S) and deposition (D) at meander bends. The channel shifts continually, leaving scars of former positions and oxbow lakes not yet filled with sediment.

62. Skillet Fork in Wayne County had a naturally meandering course across a very flat plain. It was first leveed and more recently the channel was trenched and straightened.

63. The Albion NW topographic quadrangle shows the meandering course of the Little Wabash River with many oxbow lakes marking abandoned channels. Rattlesnake Bluff is the margin of the upland where river erosion is cutting laterally against it.

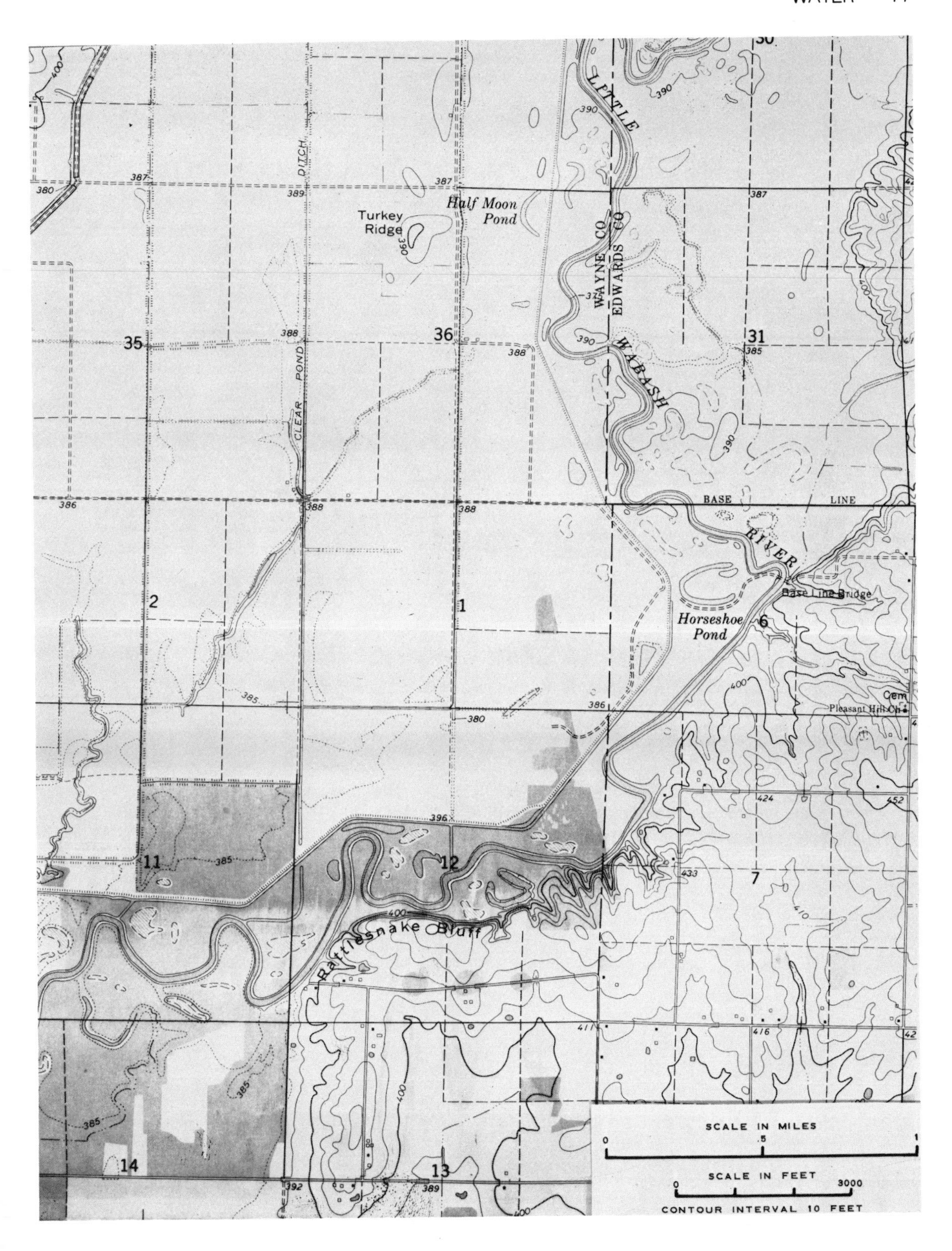
Half Moon Pond
Turkey Ridge
DITCH
CLEAR POND
LITTLE
WAYNE CO
EDWARDS CO
WABASH
RIVER
BASE LINE
Base Line Bridge
Horseshoe Pond
Pleasant Hill Ch
Cem
Rattlesnake Bluff
SCALE IN MILES
SCALE IN FEET
CONTOUR INTERVAL 10 FEET

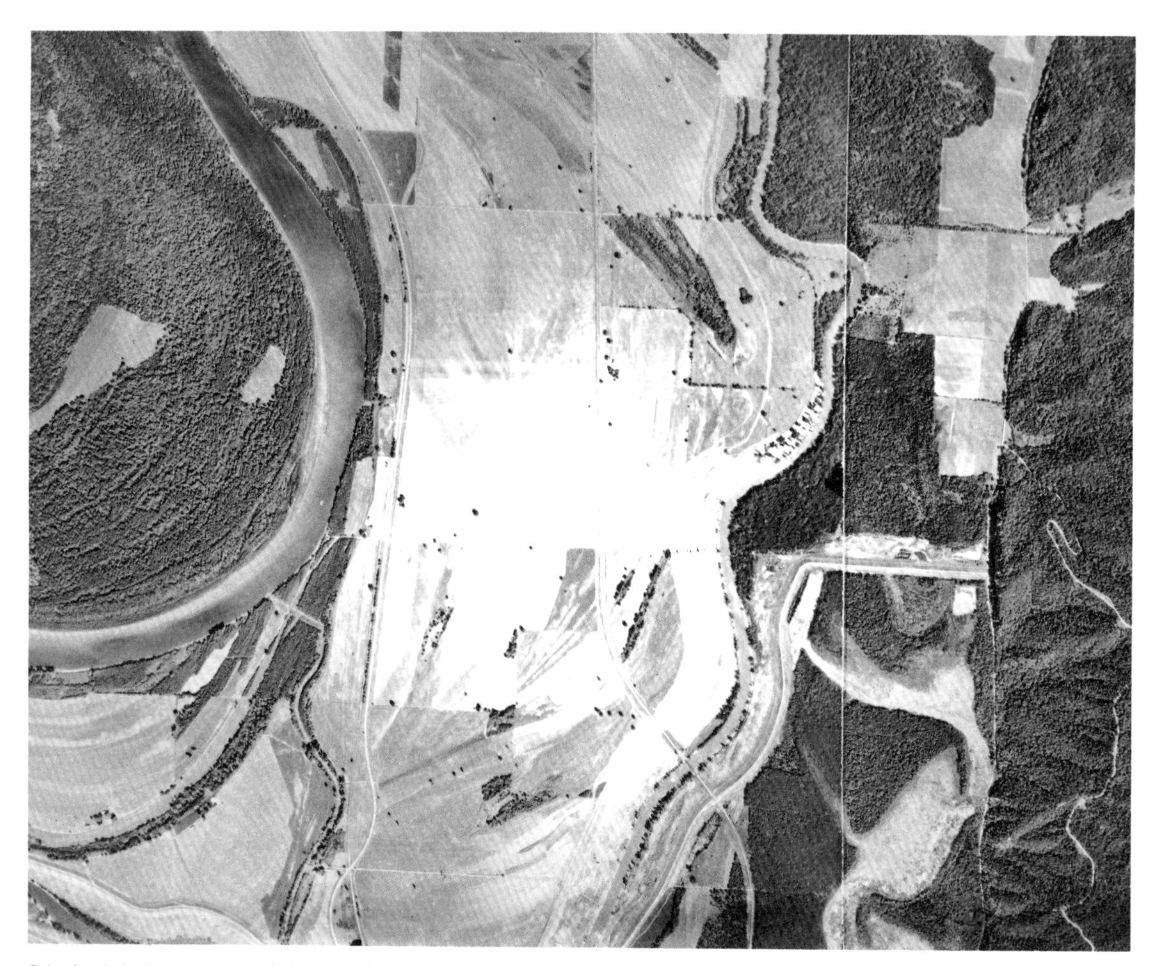

64. Aerial photograph of Tower Island Chute of the Mississippi River showing its intersection with Big Muddy River. The former meandering course of Big Muddy is clearly seen in the forest swamp at the southeast part of the photograph. (The straight white line at the right is the junction of two photos.)

Terraces

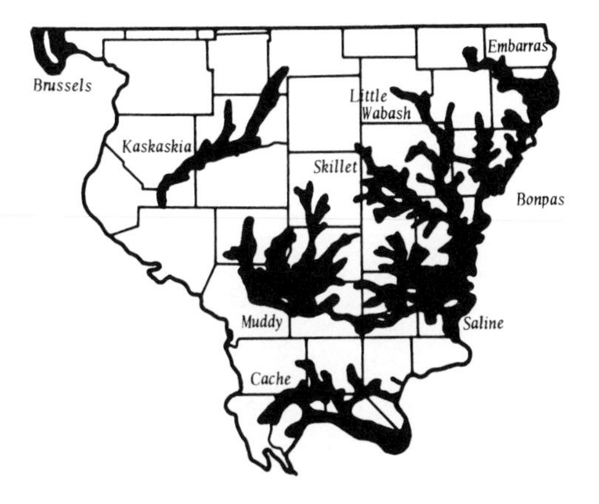

Murphysboro 1:24,000

Every stream system in our area has terraces. Terraces hold neither the scenic attraction of the bluffs nor the hazards of the floodplain. They may be valuable real estate and they do hold useful keys to the geologic history of the recent past. They are flat landforms, associated with bottomlands, in places somewhat dissected by low-order waterways. They are a step above the flat floodplain and are often set off from it by a sharp rise. Two terrace levels are clearly visible in the downstream portions of the tributary streams to the Mississippi, Ohio, and Wabash rivers.

Our terraces are of two types. One is a flat bench of ancient river sediments uneroded by the river as it deepened the rest of the valley bottom. Such terraces slope down valley approximately parallel to the new floodplain. The second is a complex of lake sediments deposited in more or less quiet water and having a nearly horizontal surface.

Whenever you cross a stream with a flat bottomland look for terraces. Modern highways cross the floodplains on a fill above expected flood levels. As you drive, the fill suddenly ceases and the highway continues level with the land. This is the terrace. Soon an abrupt or gentle rise takes you from the terrace to the upland (Fig. 65).

Larger terraces are easily identified on the 7½ minute topographic maps. They show up as flat surfaces set off from the floodplain and upland by two, three, or more, closely spaced contour lines. Murphysboro in Jackson County is built on two separate lake terraces (Fig. 66). Carmi and Mt. Carmel are on the large terrace areas which parallel the Wabash River. Soils maps delineate terraces well because certain soils types have developed in response to their particular environment (Fig. 67).

To see what a terrace is made of look for a place where a river bend is now scouring against it. A recent collapse reveals a section of layered sand, silt, or clay. If the moisture content is just right the damp coarser layers enhance the contrast. The material can be easily dug away with a shovel, for it is unconsolidated and without cementing material. On the higher terrace a deposit of well-sorted and unstratified windblown silt (loess) overlies the water-laid sediment. The effect of weathering and soil formation has altered the upper part of the section. Man-made exposures such as a sand and gravel pit, an excavation for a building foundation, sanitary landfill, or highway road cut, provide better exposures than nature does.

Poorly drained areas cause swamps and marshes. Terraces are flat and underlain by dense soil that allows only slow percolation of water. Consequently during the wet seasons or at times of heavy rains they are subject to flooding. In the lower basin the terraces are well above river flood level.

The elevation of the lake terraces remain nearly constant or actually decline farther up the rivers, especially in the Big Muddy Basin, while the floodplain levels gradually rise upstream; in mid-basin the wide terraces and

65. Level terrace in the middle ground was once the floodplain of Clear Creek in Union County. Most of that old floodplain has been eroded away, leaving a few terrace remnants. The slope between terrace surface and the present floodplain shows bare patches resulting from modern sheetwash.

floodplains coalesce, causing natural swamps which persist the year round.

Drainage ditches have been cut into many terrace surfaces in an attempt to remove surface and soil water more quickly. Generally straight features, the banks tend to slump in and vegetation will fill them unless continually kept clear. The risk is too much erosion.

Sand dunes

Sand dunes are associated with the sandy terraces of the Wabash Valley (Fig. *68*). These are low mounds or ridges which rise above the general landscape level. Many houses have been built on them, and road cuts reveal their makeup. The sand is stabilized and soils have formed in them. However, cultivation loosens the sand, and bare fields are subject to blowing.

Origin of the lake terraces

During periods of glacial advance floods of meltwater carrying vast loads of sediments coursed down the Mississippi and Wabash valleys. During the Wisconsinan stage, some 20,000 years ago, great valley trains of sediment filled the valleys and the level of the floodplain by more than two hundred feet. Not so in the Big Muddy Basin and other valleys whose headwaters had not been reached by the glaciers. Instead, backwater from the main rivers caused lakes to form. The lake level must

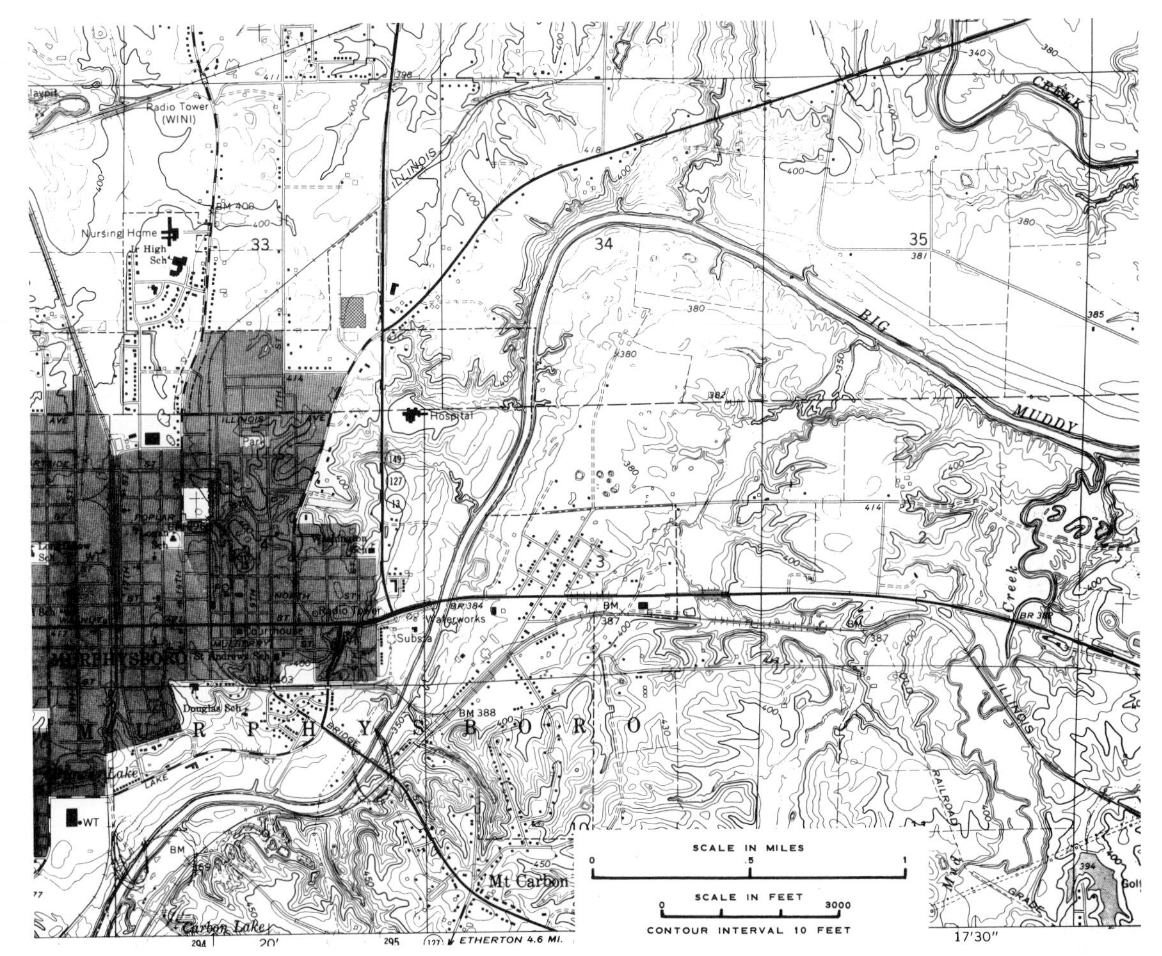

66. *Murphysboro terrace surface sloping upstream on Big Muddy River and Beaucoup Creek. The slope of the terrace north of the river is toward the north and northeast.*

have risen in summer and fallen in winter. At first flood in the spring, the meltwaters flowed back into the tributaries and built deltas and a sloping terrace upstream. This phenomenon is most marked in the Big Muddy Basin where Murphysboro is built on a considerable delta (Fig. 66).

Features to look for. If you are lucky and have sharp eyes you may find the fossil shells of small snails and tiny Ostracoda. They are preserved in silts and clays of the lake beds. Their discovery is more likely if the deposits are calcareous (will fizz when a drop of dilute hydrochloric acid touches them). White, irregular masses of calcium carbonate concretions, reprecipitated after being leached from the overlying deposits, may also be a clue to the presence of fossils.

The sands, now lying above water level are often yellow or orange. Weathering oxidized the iron-bearing minerals giving a rusty stain to the sand grains. At certain zones the leached iron oxide has been precipitated to form spherical nodules and elongate tubes. They take weird forms.

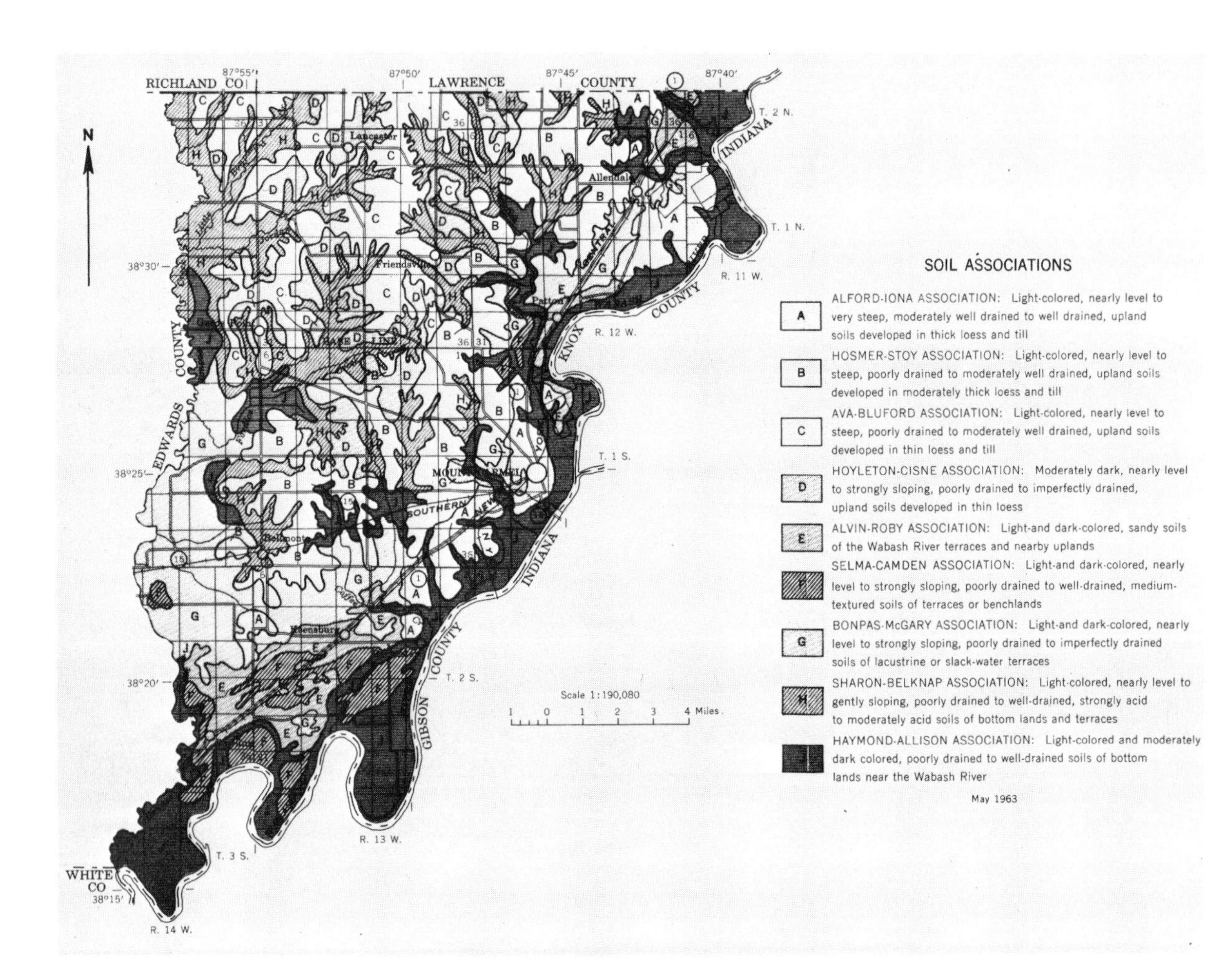

67. General Soil Map, Wabash County, Illinois

Tubes filled with a slightly different material or color are rather common. These are burrows, probably made by crayfish. They are located in an alluvial fill, the result of temporary scouring by a stream channel which was subsequently filled.

SUGGESTED READING

Frye, J. C.; Leonard, A. B.; Willman, H. B.; and Glass, H. D. 1972. *Geology and Paleontology of the Pleistocene Lake Saline, Southeastern Illinois*. Urbana: Illinois State Geological Survey, Circular no. 471. 44 pp.

68. A sand dune rising above the terrace surface east of Ridgeway, Gallatin County.

Swamps and Marshes

Karnak 1:24,000
West Frankfort 1:62,500
Vincennes 1:24,000

Swamps are widespread within the bottomland division (Fig. 69), and are found in many depressions of limestone terranes (Fig. 112). Shallow basinlike depressions were once common on the flat till plains divisions, causing wet, marshy prairies, but most have been drained.

Swamps and marshes are low places in the terrane where the water table is near or even above the surface. Open-water lakes may be associated. In fact the gradual filling of lakes with sediment and organic matter causes the transition to swamp. Such places are most favorable for living things; they are environments where the earth scientist and the biologist meet. Many of the physical conditions are the results of biological activities.

The mud and sediments beneath the swamp may tell a fascinating story of changing conditions. The sediment is generally fine-grained silt and clay but a layer of sand tells of floodwaters. Fossil remains of plants, pollen, shells, and bones tell us about past inhabitants and perhaps climatic change.

A biologist would concentrate on the plant and animal inhabitants, but the physical conditions are interesting too. A geologist would ask "Why is it here?" and a conservationist would ask "Is it safe from destruction?" A topographic map and aerial photograph would be most helpful because flat land, wet ground, and thick vegetation make exploration difficult. These maps will show the shape and the boundaries, whether a depression on a flat surface or in irregular ground. Black Slough in Johnson County (Fig. *44*) is surrounded by high hills. Water drains slowly to Cache River which itself is winding and slow moving. The swamps date back to glacial times when the ancient valley of the Ohio River was close by. Glacial sands and gravel filled up its valley and caused all the tributaries to fill up as well. At first there were lakes but gradually they filled with sediment. A visit to the Heron Pond Nature Preserve today will show that the channel of the Cache has eroded fifteen feet beneath the level of the swamps. Surface drainage into the channel is causing gullies, and animal burrows cause underground "piping." Post Creek cutoff shortened the length of the river by sixty miles, and a steeper gradient permits deepening. As a result many old swamps of the main valley have now drained so as to permit cultivation, but gullying continues.

Where a watercourse enters a swamp, sediment drops out. Look for evidence of delta building. A flood may carry coarse material well into the swamps. How far can a waterway

69. LaRue Swamp from Pine Hills, Union County. The winding course of an abandoned channel of Big Muddy River contains both swamp and open-water ponds, at the margin of the Mississippi River floodplain.

be traced through the swamp? Is a single channel evident?

Many swamps are, at least partly, due to the presence of springs. The significance of the springs may be evident only in dry seasons or even during a drought when the area stays wet and the location of the spring outlets can be identified. Fluctuation of the water level is characteristic and dry periods bring stress to the life of the swamp. The vegetation may even burn at such a time, leaving ash and charcoal which may be discovered in an auger boring or core hole.

Most sinkhole swamps are rather small. Furthermore, their life expectancy is short. When the organic debris clogging the bottom of the sinkhole gives way, the water drains out and the swamp is no more.

Marshes are dominated by grass, sedge, and reed rather than arboreal vegetation. Those known to us are near streams on bottomlands in both the plains divisions and the bottomlands divisions. The reasons for the difference in vegetation are not known. Perhaps drying followed by fire is one factor.

Marshes were probably rather common in the prairies under natural conditions, but they were drained and cultivated. The geologic map of the Belleville-Breese area, dated 1915, shows many marshy areas.

Natural Lakes and Ponds

Thebes 1:62,500
Thebes 1:24,000

Natural lakes and ponds are found throughout the bottomland divisions and in the limestone terranes of the Ozark and Shawnee Hills divisions, though our area is not particularly noted for them. Most are abandoned river channels with elongate and arcuate shapes (Fig. *70*). In recent years the protected beaver have built dams across a number of bottomland watercourses. Such structures tend to be only temporary; they are abandoned when food is depleted or when the Conservation Department relocates the colony. A third type of pond is oval and found in clusters on the limestone terranes (Figs. *111, 112, 113*). Many of the natural lakes have been modified by dam structures to increase the size of the water body and to regulate the water level.

Geologically lakes may be formed by every kind of surface and many subsurface processes—any process which will produce a depression or raise a dam. Landslides may dam a mountain valley, and depressions form between sand dunes or irregular glacial deposits. Such conditions don't exist in Southern Illinois, although once the morainal surfaces must have had many kettle lakes. Illinoian glacial ice formed lakes by damming the north-flowing streams of the Shawnee Hills. Remnants of their deposits are hard to find. The extensive terraces of the Big Muddy and Saline

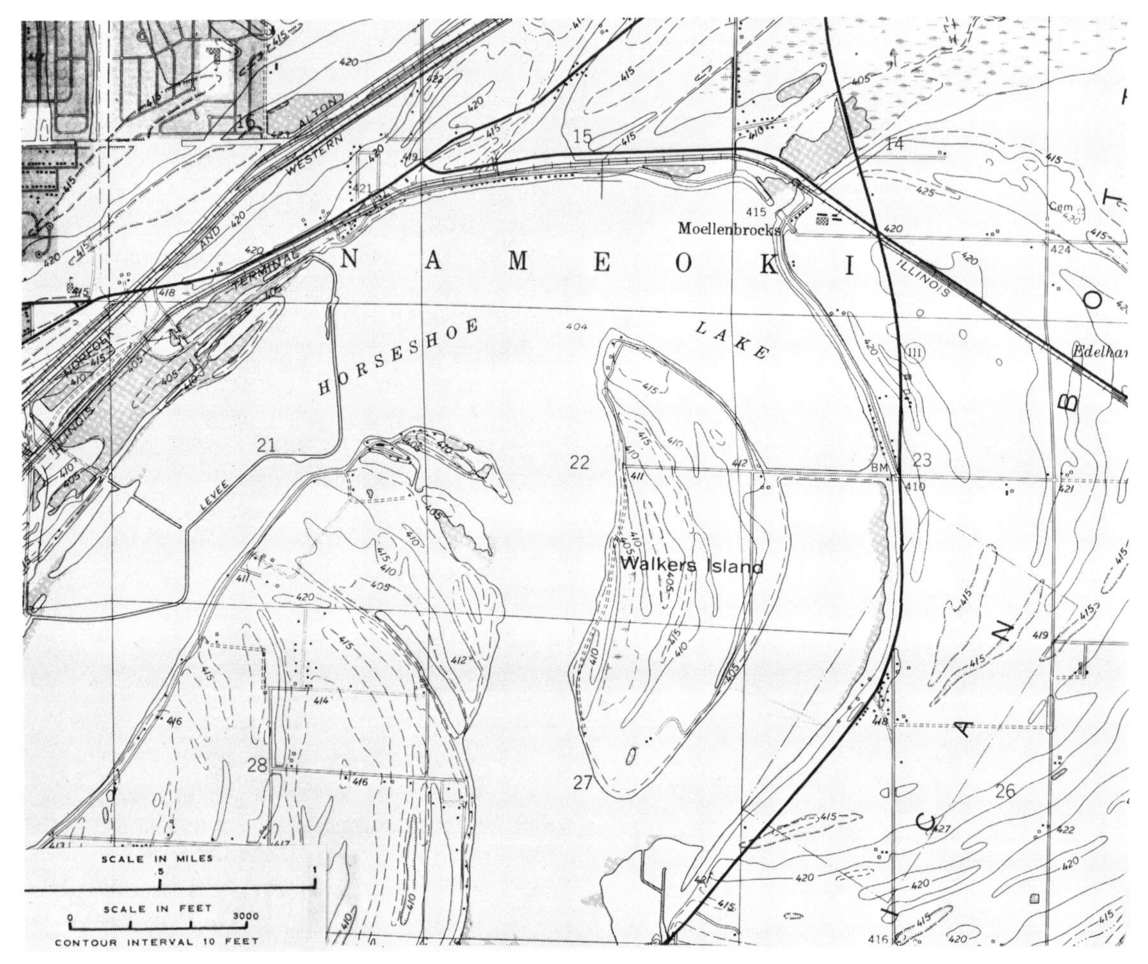

70. Horseshoe Lake on the American Bottoms, St. Clair County. The shape of the lake and the elongate ridges and swales establish this as a former channel of the Mississippi River.

valleys are lake-floor deposits caused indirectly by the later Wisconsinan glaciation. The lakes drained away when the Mississippi River deepened its valley.

Lakes are short-lived geologic phenomena. Their basins become filled with sediments or their outlets are deepened and the lake is drained. Our bottomland lakes are gradually filling with sediment brought during floods and with the organic matter which grows along the shore and out into the water itself. The ponds of the limestone areas may also fill and become marshy basins, but more often their water may suddenly disappear underground as the plug of sediment and rotting vegetation gives way and releases the water into the cave system below.

Bottomland lakes are depressions below the water table. Most are shallow, but oxbows of the bigger rivers may be twenty or thirty feet deep. Once the main river flowed through this channel (Fig. 71). Its turbulence scoured deep around the bend, but a new course cut a shorter channel and silt dropped out into the quieter

71. Round Pond in Gallatin County, an abandoned segment of the Ohio River channel against the Shawneetown Hills. Ohio River and Shawneetown Bridge in the background.

waters of the bend. At last the ends of the loop became closed so that only at floodtime does the river enter the oxbow.

Perhaps these lakes are most noteworthy as a favorable habitat for life. Deep water and shallows, a dependable water supply even in the dry summer, support an abundance, and a wide variety, of plants and animals. These are places to go to fish, to enjoy nature, to search for birds, reptiles, and amphibians, and to study the association of water and land.

Changes take place rapidly even under natural conditions, but man's activities rapidly alter the vegetation, change the water level, or the natural seasonal cycle. We have little concept of what these lakes were like 150 years ago. Aerial photos between 1920s and the present reveal many changes even in the recent past.

Sinkhole ponds are features of the upland (see Figs. *112* and *113*). Most have neither surface inlet nor outlet; some have a small drainage area. The outlet is underground through slow seepage. Some ponds are therefore seasonal.

Sinkhole ponds are found in clusters associated with depressions which don't hold water. Some depressions have open cavities in the bottom, leading to large underground rooms or cave systems. When such openings become clogged with tree branches, leaves, and other debris they may hold water for a time. Subsequently eroded soil may seal the bottom, and a pond is retained. Several such ponds are known above Southern Illinois caves. A continual drip of water has produced stalactites and flowstones.

The seal of debris in the bottom generally lies well above the water table. Solution cavities below lead to a surface-stream valley, perhaps several miles away. The seal is thus temporary and may gradually, or even suddenly, give way and the pond is drained. Residents of Hardin County tell of the time the pond in Big Sink disappeared overnight. This was a large pond covering at least forty acres. Many people came with wagons and trucks to collect the haul of stranded fish. A search along the Ohio River for the outlet of the underground passage was unsuccessful. Subsequently the outlet became sealed again and the pond was used by the Minerva Fluospar Company to supply processing water.

SUGGESTED READING

Russell, I. C. 1895. *Lakes of North America*. Boston and London: Ginn and Co. 125 pp.

Coker, R. E. 1954. *Streams, Lakes, Ponds*. Chapel Hill: Univ. of North Carolina Press. 327 pp. (Reprint. New York: Harper & Row, Torchbooks, 1968.)

Man-Made Lakes and Ponds

Burnt Prairie 1:24,000
Crab Orchard 1:24,000

Lakes and ponds are everywhere. Every modern topographic maps shows farm ponds, town reservoirs, recreation lakes and, most recently, waste-disposal lagoons and holding basins. Big lakes made by damming the larger streams extend across more than a single map. Important as lakes are to supply the water needed during the dry season, much public resistance is building against the schemes to construct hundreds of additional dams.

Ponds on topographic maps are mostly in low-order drainageways, but sometimes simply in a basin excavated on the hillslope. Enough water flows off the upland to fill them in this position. Dams built across higher-order valleys are more expensive to build, present greater problems of seepage below the dam, and require elaborate emergency spillways to protect them against erosion during floodtime.

A pond is well worth visiting. Pond life is fascinating, but beyond that notice the topography around it. Most ponds are on the hillslope at the beginning of concentrated flow. See if the dam is built across a waterway or on the slope where runoff has not been concentrated. The reservoir may have been partially excavated so that the divisions of the loess and perhaps pebbly glacial till can be clearly seen. An old pond may have some shoreline erosion and a bank at the edge. The larger the pond the bigger the waves and the more the erosion. Look for deltas of sediment washed into the pond by streams or gullying. If there has been excavation, note whether there are bare areas suffering erosion and resistant to the growth of grass cover.

The spillway of the dam is critical because overflow at times of very heavy rain (Fig. 72) may cause erosion. Seepage of water beneath the dam is very common. The grass is more lush below the dam, and there may even be visible seepage. Burrowing animals such as muskrats and crayfish can weaken a dam and piping of water through the burrows leads to trouble. Sewage lagoons and holding ponds

72. *Spillway of Crab Orchard Lake, Williamson County, was placed at the valley side so as to take advantage of resistant sandstone. Considerable erosion has taken place since 1940.*

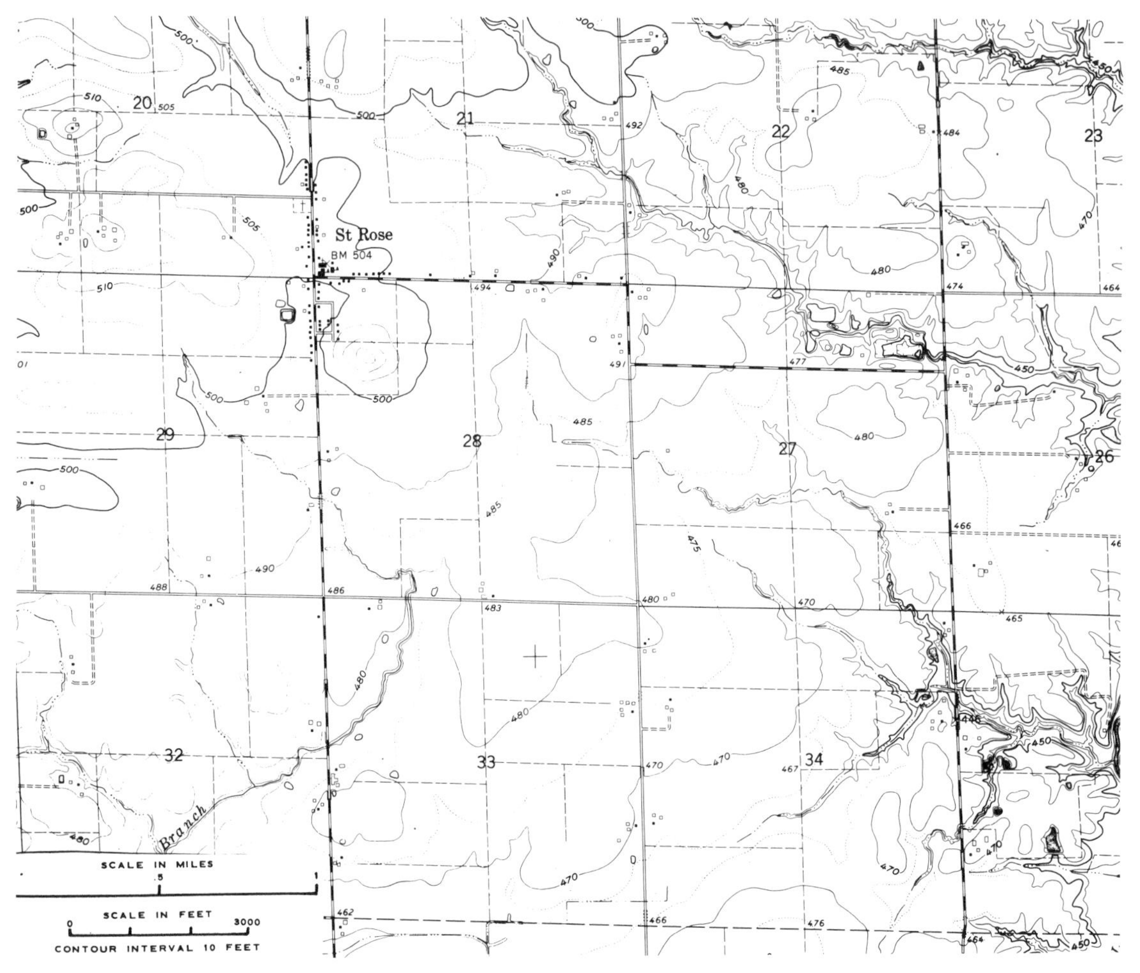

73. Part of St. Rose topographic map showing many small ponds in the low-order drainageways. Note the rectangular disposal lagoon near St. Rose, Clinton County.

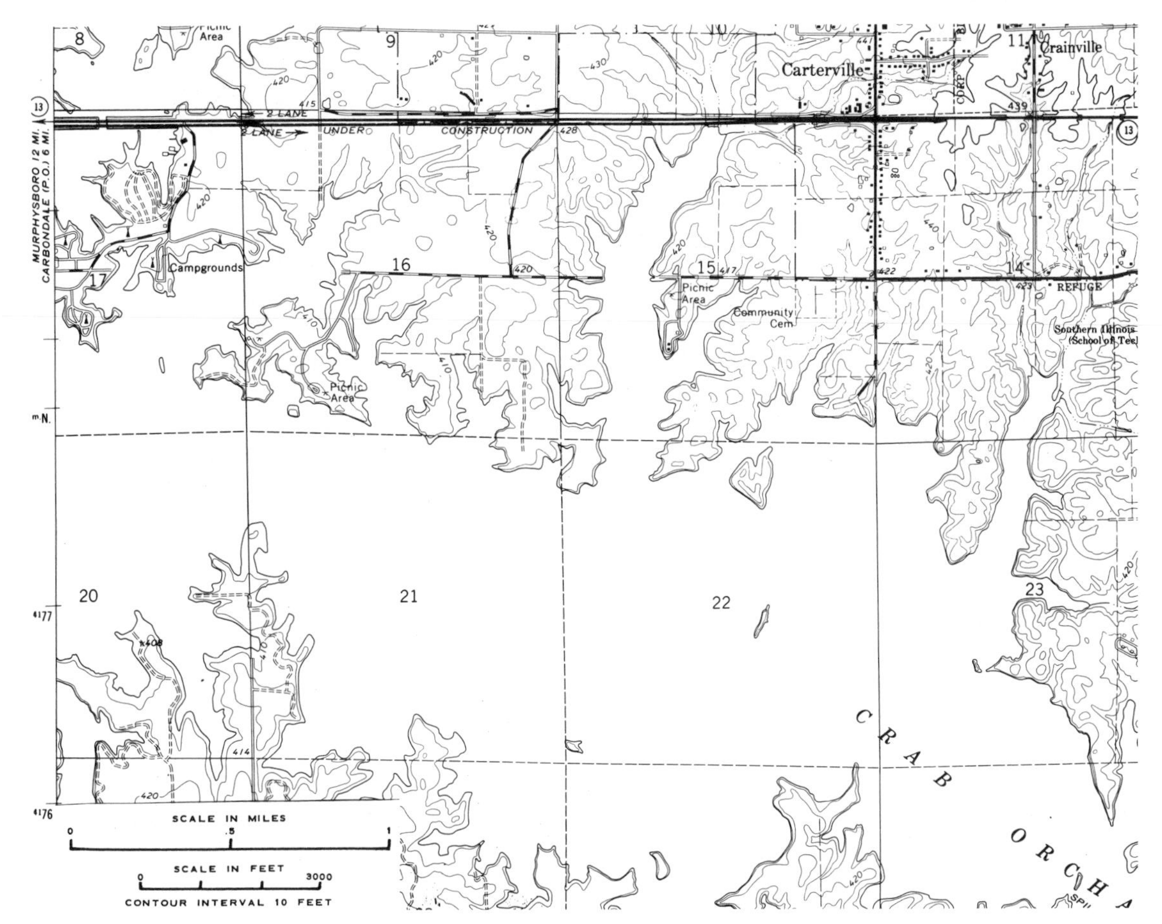

74. Crab Orchard Lake, Williamson County, 1:24,000. Man-made lakes, formed by damming a large stream, have a long, irregular shoreline because the lower course of each tributary stream is also drowned.

75. Shoreline of Crab Orchard Lake, Williamson County. The cliffs, formed by wave erosion, are of loess and glacial till. The shoreline and lake floor are muddy because there is little naturally available sand.

76. Devil's Kitchen Lake. Steep hills and sandstone cliffs border the narrow man-made lakes of the Shawnee Hills Division.

commonly have more regular outlines and a dike may be built above the general ground surface (Fig. *73*). Water in these ponds is being purified through settling and precipitation of solids, and biological alteration of wastes. Waste waters may be led through a succession of ponds, each one being of higher quality.

The larger lakes such as Carlyle and Little Grassy display a wide range of interesting features. Each is distinctive and each is undergoing dynamic changes in shoreline patterns (Fig. *74*). Compare the shoreline of a new lake and one which has existed for a number of years. In the summer of 1973 the water surface of Rend Lake gently met the enclosing hillslopes. We can predict that in the year 2,000 the shoreline will be marked by cliffs similar to those on Crab Orchard Lake (Fig. *75*). The headwaters of these lakes are very shallow. Extensive accumulations of sediment are deposited at the mouth of each tributary stream, and the fine silt is carried out into the lake basins.

The silty loess which covers the upland of Southern Illinois has a profound effect on our lakes. The shorelines lie against the silty soil which is easily eroded by wave action. Such material has no sand and no pebbles, but some clay. Eroded clay and fine silt stay in suspension and make the lake waters murky. No coarse material is present to lag behind and produce a nice beach. Instead the beach is mud. Glacial deposits are being eroded along much of the shoreline of Crab Orchard Lake but that material also has little coarse material. A few pebbles and boulders line the shore but the sand grains are lost in the dominating silt. The sand at public beaches has been trucked at considerable expense from one of the major rivers.

Lakes in the Shawnee Hills have steep shorelines because the dams were placed to take advantage of high cliffs and a narrow valley. Devil's Kitchen (Fig. *76*) and Little Grassy lakes are ninety feet deep at the dam, and sandstone cliffs characterize the area. The ridges and valley slopes have a thick cover of loess. Thus, runoff brings silt to these lakes too. Theoretically some sandy beaches might form here, but the water is deep next to the sandstone cliffs, and the shoreline of the shallower coves contain more silt than sand. The rocky tributaries supply coarse sandstone cobbles, but they, too, are lost in the lake.

Intensive use of the hillslopes near the lakes in the Shawnee Hills is hazardous because the soils are easily eroded when exposed. Secondly, the soil does not offer good infiltration properties for the proper disposal of sewage. Near the lakes the bare rock forces effluent to the surface where it washes into the lakes. The developed areas of Lake of Egypt showed algal blooms in 1973, and danger of severe pollution is imminent.

SUGGESTED READING

Roberts, W. J. et al. 1957. *Potential Water Resources of Southern Illinois*. Urbana: Illinois State Water Survey, Reports of Investigations no. 31. 100 pp.

Smith, W. C. 1966. *Geologic Factors in Dam and Reservoir Planning*. Urbana: Illinois State Geological Survey, Environmental Geologic Notes no. 13. 10 pp.

Upland Landforms

The Upland

Illinois is the prairie state. Flat or gently rolling prairies extend as far as the eye can see. We have segmented it into great cultivated squares (see Fig. 77). Stream are widely spaced and the upland is poorly drained. Central Illinois is flatter than our lands south of Interstate 70. In the Central Plains part of Southern Illinois every topographic maps shows streams and small headwaters dissecting the prairies (Fig. *23*). Early settlers must have found few upland depressions and marshes, though the widest flatlands such as Twelve Mile Prairie were certainly very wet in the springtime. Today waterways are commonly bordered by trees (Figs. *78* and *79*), perhaps much as the pioneers found them.

The uplands provide most of the agricultural land and desirable sites for towns, for industry, and transportation routes. Flooding is not the problem, so characteristic of river bottoms, but on flat divides water may stand in the fields after long periods of rain or during the spring thaw. Many fields have been tiled to speed the drainage. Soils developed on flatlands have tight subsoil which inhibit infiltration and underground movement of water (see sec. Soils).

Southward in the Shawnee Hills (Fig. *87*) and the Ozarks the uplands are hills and ridges rising two hundred to five hundred feet above the valleys. Their forms and contours reflect the patterns and resistances of the rock beneath them. Rocks are exposed in high bluffs and in man-made road cuts, quarries, and mines.

Rainfall on the upland does not immediately run off. It fills every small depression, is held around and between plants, all the time soaking into the ground. Eventually water flows from depression to depression and then between plant stems or litter as a film or sheet. Little erosion occurs unless the ground has been bared or cultivated. On the slopes rills and gullies may erode formerly undissected surfaces, especially in fields and pastures.

It might seem that the uplands offer less of interest than the valleys. That is not so. High places offer scenic vistas, and the ground beneath offers materials to collect and a challenge to interpret its history. Soils have infinite variety, yet enough unity so that they can be classified on the basis of their properties. Pamphlets and maps obtainable from the Soil Conservation Service will yield days of adventure. Soils were formed from the weathering of windblown silts, glacial deposits, and bedrock. Distributions of the latter are shown on geologic maps. Figure 77 shows the marginal boundaries of glacial deposits, coal beds, and some thin limestones in the plains division. Features to look for are described in the sections which follow.

Hillside Scarps

Hillside scarps are conspicuous erosional features in many upland areas, especially in hilly pastures which have been overgrazed. The scarp takes the form of a low cliff facing downhill (see Fig. *80*). The scarps are most commonly a foot or two in height but may reach fifteen feet. They are formed by rapid sheet and rill erosion of the soil's parent loess or glacial till. The scarps are crenulated by basinlike reentrants where erosion proceeds in a small amphitheater, and by promontories where erosion is less rapid.

The position of the scarp on the hillslope is at the edge of the ridge crest where the gradient steepens. Some scarps extend for many

77. Flat prairie southeast of Nashville, Washington County. This is the divide between drainage to the Kaskaskia River (toward upper right) and Beaucoup Creek (straight ditch in left foreground). The three white lines in foreground mark a pipeline.

78. Aerial photograph of Jefferson County. Trees mark the drainage lines. Light-colored areas mark slopes where sheet erosion is scalping the soil.

79. Rolling upland west of Campbell Hill, Jackson County. Fields are cultivated, but the wide swale of low-order waterways has been left in grass. There is no gullying. Trees occupy the valleys.

80. Hillside scarp, an active erosional feature visible in many pastures and abandoned fields. Sheetwash, rain impact, mud slides undercut the sod and cause retreat of the scarp. Gullies are rarely associated with it.

yards, others are discontinuous but in the same topographic position. Gullying is seldom found on the upland above the scarp nor do gullies develop from them. The scarps are formed and grow through erosion by rain impact, sheetwash, and slumping of soil blocks. Discontinuous gullies (see Overland Flow) are associated in some places.

At the base of the scarp the ground surface flattens out, and grades into the general hillslope below. This slope is an erosional surface without any soil profile; it is like a model of one of the great pediment slopes at the foot of many mountains in the west. Hillside scarps retreat rapidly and become higher as they erode into the hill. A succession of photos taken over a year's time would record rapid changes and reveal a number of processes at work.

These features are nature's response to changes in the vegetative cover and the intensity of man's use of the land. They are insidious because the entire soil profile is removed—in some places the entire thickness of the loess. On the upland ridges south of Makanda, Jackson County, the scarps have eroded down to the bare sandstone and the entire mantle of unconsolidated materials is being removed.

Soils

Soil is familiar to us, yet also mysterious. The farmer knows it well, for he tills it and plants his crops in it and harvests the produce which sustains us. He knows it must be worked at the proper moisture content, that storms cause wash and gullying, that organic matter improves the tilth, that fertility must be maintained by returning or adding nutrients. Soil properties are critical, too, for construction, for disposal of waste, for recreation—indeed for any intensive use of the land.

Soils have received more intensive study than any other earth material, even more than metals, yet few of us know well even the soil in our front yards. Agronomists have described the soils of Illinois, and every county has a soil conservation office where information about soils and help regarding the use of specific soils can be obtained. All counties have been mapped, although the detail and recency vary greatly. Careful mapping has been done in the conservancy districts and in the project areas for the large artificial lakes.

Soils are most visible in the Central Lowland Division where the land is flat or gently rolling and widely cultivated. In the hilly provinces most of the land is clothed by vegetation, and soils are not so easily observed. Where slopes are laid bare we are likely to see the ravages of rain and frost upon a soil nurtured beneath a cover of vegetation.

Soil is a product of physical, chemical, and biological alteration of the earth materials lying at the surface of the ground. Soils are dynamic and gradually change their character when subjected to cultivation. Some soils are mature or old in the sense that the parent materials have been greatly altered and are close to equilibrium with the environment. Other soils are immature in that new material is continually being added by wind or river deposit—or erosion is proceeding so rapidly (Fig. *81*) that weathered material is carried away as fast as it forms.

Soil is three dimensional. The surface horizon is most directly affected by vegetation, microorganisms, burrowing animals, and cultivation, but roots of some plants penetrate thirty feet and more and animals burrow several feet. Physical and chemical processes are most intense at the surface, but water and air circulate through the soil. The capacity of a soil to take in water and transmit it is a prime attribute. Temperature changes lag behind the season as soils and rocks are poor heat conductors; temperature ranges decreases with depth until at some twenty feet temperature stays the same as the average annual temperature of the area.

Color is the most obvious and useful property. Wetness tends to make color brighter and helps distinguish subtle changes in tone and hue. Color is so important in soils work that professional soils people carry color charts for use in classification. Dark tones of gray or brown indicate the presence of organic matter. The brown and buff soils of hilly lands were formed under forest cover. Some subsoils are mottled brown and gray, and change in color with depth indicates different horizons with different properties. Black soils are rich in humus.

Texture is a measure of the size of particles. Soil descriptions include a terminology based on the percentages of clay, silt, sand, and pebbles. A good field test is to squeeze a bit of damp soil between the fingers. Sand is very gritty, the individual grains can be easily seen and they will not stay together when unsupported. Silt is stickier. A lump squeezed between the fingers will feel smooth. It can be squeezed out into a short ribbon but the surface looks textured and the ribbon breaks be-

81. Surface of a steep slope in the Shawnee Hills where most of the surficial material has been eroded away and shale fragments of the bedrock are exposed. There is no longer a true soil here.

fore it gets very long. Put a little between your teeth and you can feel the gritty particles. Clay is very sticky when wet, but hard as a rock when dry. Damp clay can be pushed out into a long ribbon which doesn't break apart. When rubbed hard it takes a polish, and between the teeth it feels smooth as butter.

Most soils include a wide range of textures, and "loam" is a mixture of silt, sand, and clay, plus organic matter to give a texture which is most favorable to plant growth and to man's activities. Stony soils are present in rocky hill lands.

Structure is most clearly visible when soils are dry. The soil mass contracts as it dries and tends to break into flakes or blocks or loose aggregates of different sizes and shapes.

Texture and structure together determine the capacity of the soil to transmit water and air, the ease of root penetration, and, in some measure, the chemical activity within the soil.

A chemical measure of soil is the pH. You cannot see the pH, but with experience it can be estimated indirectly. The species of plants which grow best or do poorly on it are good indicators. Certain structures of the soil and secondary precipitates are also indicative. The pH is a measure of the hydrogen-ion dominance. The more dominant, the more acid. Alkaline soils have a dominance of calcium, magnesium, potassium, or sodium. In Southern Illinois the more mature soils on the upland tend to be acidic, that is, they have a low pH, whereas more recently formed soils and the grassland soils tend to have a nearly neutral or somewhat alkaline pH.

82. Profile of a soil exposed in a road ditch west of Carbondale, Jackson County. The A horizon is about six inches thick; it has been considerably disturbed. The B horizon has a blocky structure, very dense in the upper part. It has a lighter tone in the lower part. The profile is complicated here because weathering extends from loess above into glacial till.

Soil Profile

A ditch or trench often exposes the soil profile which consists of two or more horizons downward from the surface. These do not represent sedimentary layering. They are the result of chemical and physical alterations brought about by weathering processes. Figure *82* is a photograph of a soil profile showing the main horizons.

The surface soil is technically called the A horizon. It is most directly affected by vegetation, microorganisms, burrowing animals, and cultivation. The A horizon is generally darkest and has most organic matter. Physical and chemical processes are most intense near the surface, air circulates freely, and rainwater penetrates and sinks downward through it. As a result soluble substances have been lost. Clay particles may also have been carried downward into the B horizon.

The subsoil, or B horizon, is generally tighter,

containing more clay and secondary concretions of the leached materials from the A horizon. The structure is often blocky when dry and sticky and tight when wet (Fig. 83). The prairie soils of Southern Illinois are notorious for their "claypan" which inhibits the percolation of water. Black, "buckshot" concretions of iron-manganese oxide are common in several soils series. Where sheetwash and cultivation expose the B horizon these small spheres are prominant on the surface. Brown and rust-colored mottling and streaks are also common. In some of the flat prairie lands "slick spots" on the surface locate places where sodium compounds have accumulated in the subsoil.

The contact between the B and C horizons may be irregular or gradational. The C horizon is generally thickest and extends downward into unaltered parent material. It is generally oxidized, may contain secondary material but has lesser soil structure and more features of the parent material.

Soil profiles in Illinois are generally compound and two or more distinct soils, one above the other, may be recognized. This is because the ancient bedrock soil was covered by glacial deposits on which a second soil developed. Subsequently, layers of windblown loess were deposited on the glacial deposits and soils formed again. In Madison County

83. A road cut in loess, Jackson County. The dense sod holds the A horizon together, but the B and C horizons are more susceptible to disintegration and erosion. Sassafrass roots penetrate to considerable depth but do not hold the subsoil.

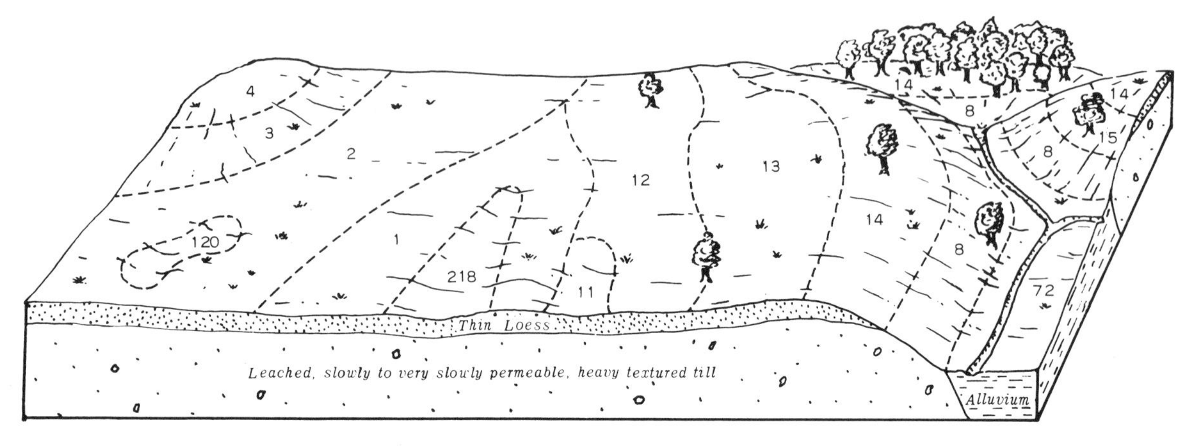

RELATIONSHIP OF SOILS ASSOCIATED WITH THE CISNE SERIES

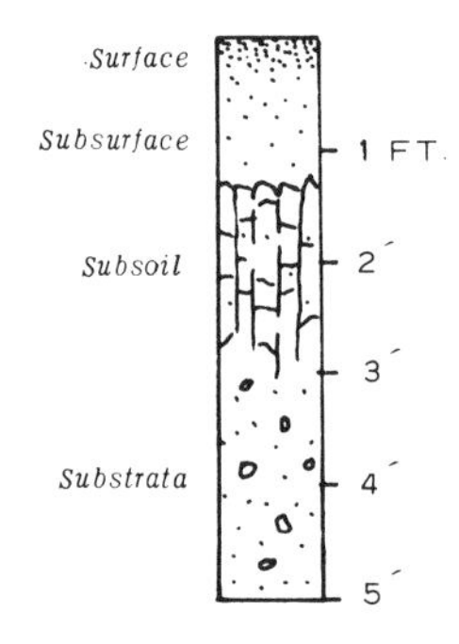

84. Soils develop different properties under the same climatic conditions, depending on the nature of the underlying material and the surface slope. The diagram represents the Cisne Series (2) and related soils. The profile is representative of Cisne Series features.

dark soil zones between distinct loess deposits can be seen in many places.

Soil Maps

General soils maps showing distribution of soil associations are available for most counties in Southern Illinois. A brief description of topographic aspect of each soils series in the association and the properties of the major soil series are given. The map (Fig. 67) shows the soils association of Wabash County. Figure *84* shows diagrammatically the topographic relationships of the Cisne soil series and associated soils. A few county reports have been published in the recent past and several others will soon be released.

Soils series represent the basic mappable unit; each is given a specific number. Each series has a set of properties which distinguishes it from other soils. Soils within a series may differ somewhat in the texture at the surface or the slope or other aspect affecting its use by man, and are considered separate phases identified on maps with a letter following the series number.

Soils maps are essential to modern farming and to planning for optimum land use. They are also valuable to the outdoorsman because they indicate the nature of the land, including slope, wetness, and rocky places. Areas disturbed by man, such as strip mines, quarries, and drainage ditches, are clearly located.

Things to do. From a soils map locate specific soils series and phases; visit the location and try to identify the soils. Note vegetation and human use. Examine land use to see if these are considered optimum by soils conservationists. Visit your county soils conservation office or county extension agent; you will find many useful pamphlets and people interested in your questions.

SUGGESTED READING

Fehrenbacker, J. B., Walker, G. O., Wascher, H. L. 1967. *Soils of Illinois*. Agricultural Experiment Station Bulletin no. 725. Urbana: University of Illinois, College of Agriculture. 47. pp.

Gibson, J. S., and Batten, J. W. 1970. *Soils, Their Nature, Classes, Distribution, Uses and Care*. University, Ala.: University of Alabama Press. 296 pp.

Hausenbuiller, R. L. 1972. *Soil Science*. Dubuque, Iowa: William C. Brown. 502 pp.

United States Department of Agriculture. 1972. *Soil Survey of Edwards and Richland Counties, Illinois*. Soil Conservation Service in Cooperation with Illinois Agricultural Experiment Station. Washington, D.C. 84 pp.

University of Illinois, College of Agriculture Agronomy Dept. Urbana. Detailed reports on many county soils.

Glacial Ridges

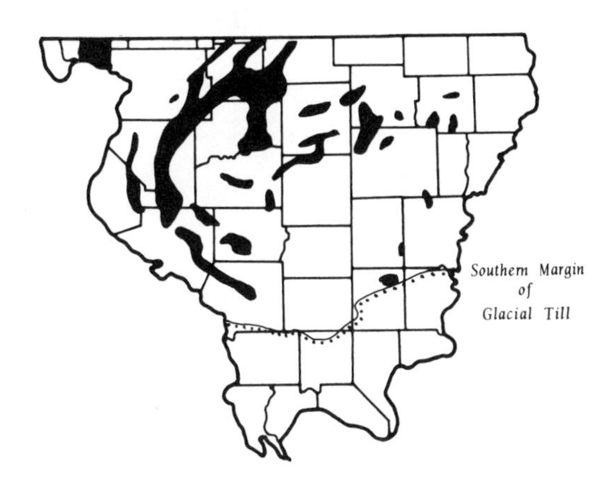

Vandalia 1:62,500
Mascoutah 1:24,000

Hills and ridges rise prominently above the generally flat Southern Till Plains Division, especially in the Kaskaskia River Basin (Fig. *85*). Some hills are conical while others are elongate, commonly in groups parallel to one another. They are composed mainly of glacial materials, some a mixture of debris, some largely of sand and gravel poorly stratified. Bedrock lies below the level of the plain and does not extend up under the hills (Fig. *86*).

Some ridges in Madison, St. Clair, and Monroe counties seem to be end moraines. The front edge of the ice must have remained there for some time—forward movement bringing more and more debris, but melting proceeding just as rapidly and releasing the debris. Other ridges, especially in Bond and Fayette counties, are lined up more nearly parallel to the direction of the ice movement. These are more sandy and gravelly. They appear to represent sand and gravel which had been carried by meltwater across or in the glacier, and deposited along crevasses in the ice. It is thought that the Illinoian ice sheet extended itself so far

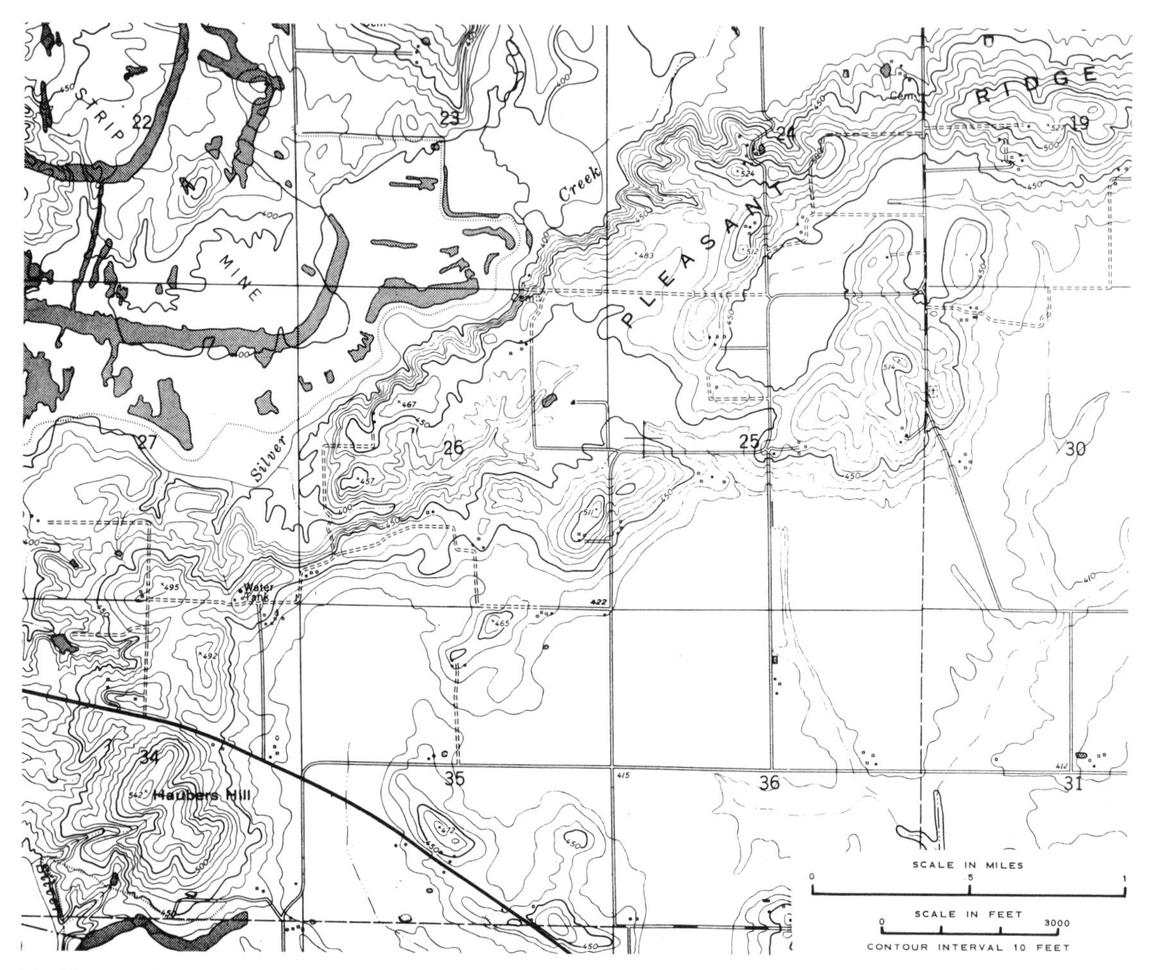

85. Pleasant Ridge, south of Mascoutah in the Mascoutah topographic quadrangle. The ridge is entirely of Illinoian glacial debris. Bedrock is below the level of the surrounding plains.

86. Glacial ridge north of Lebanon, St. Clair County. Ridges and conical hills of sand and gravel or glacial till are common in St. Clair, Bond, and Clinton counties. They rise above the general level of the upland.

from the source that as melting occurred it became thinner and thinner until forward movement ceased. It stagnated and melted down, releasing the enclosed debris. Perhaps most was carried away to the Mississippi, but many uneroded ridges remain. No sandy outwash plains exist in Southern Illinois as do those farther north associated with the later Wisconsinan glaciers.

Look for these hills above the plain. They lie between the rivers and form the high places on the plains. Generally they have rather steep slopes; most are not cultivated. Orchards cover the slopes of some and many homes take advantage of the scenic outlook these prominences provide. The glacial material is masked by windblown loess, several feet thick, deposited during Wisconsinan glaciation. Only a few sand and gravel pits are found in these hills because they apparently contain too much fine material.

SUGGESTED READING

Illinois State Geological Survey. Urbana. Write for list of Field Trip Guide Leaflets.

Jacobs, A. M. 1971. *Geology for Planning in St. Clair County, Illinois.* Urbana: Illinois State Geological Survey, Circular no. 465. 35 pp.

———, and Lineback, J. A. 1969. *Glacial Geology of the Vandalia, Illinois, Region*. Urbana: Illinois State Geological Survey, Circular no. 442. 23 pp.

Hill Lands

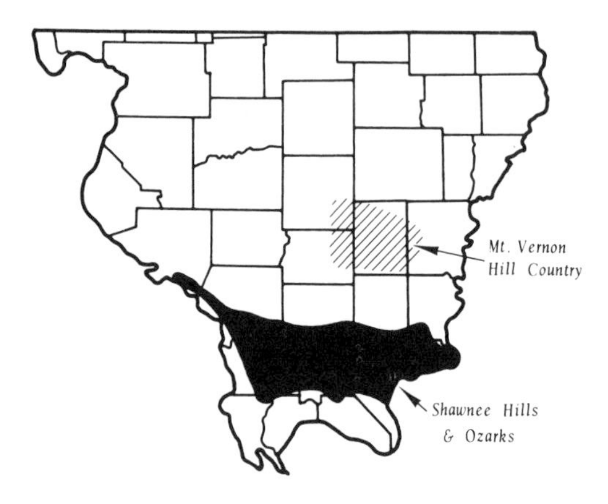

Paducah 1:250,000
Harrisburg 1:62,500
Goreville 1:24,000
Wolf Creek 1:24,000
Mt. Pleasant 1:24,000

The Shawnee Hills and the southern section of the Ozark Division are hill lands. They have relatively high relief because the rocks beneath them are more resistant to erosion than the rocks of the other divisions. Streams have eroded deep valleys with steep slopes and rocky exposures and cliffs.

The Ozark hill lands in Union and Alexander counties differ considerably from the Shawnee Hills. The bedrock is mostly chert (flint) and siliceous limestone. There are few cliffs except those bordering the Mississippi Valley and the Cache Valley on the south. Slopes are very steep and unstable. Ridge crests are narrow but covered with deep loess. Near the Mississippi Valley, source of the windblown loess, many slopes are rocky, not because of a lack of loess deposition but because of rapid erosion.

The hills and their vegetation are very interesting because steep slopes and thin soils produce extreme micro-climatic stress. South- and west-facing slopes receive the greatest solar energy. This causes rapid evaporation and very dry summer conditions, and in winter repeated freeze-thaw promotes disintegration and erosion. Examine the slopes for depth or absence of loess.

You will find many steep places where scree (loose rock fragments) covers the slopes. Search for evidence of active sliding. The scree is angular chert mostly of pebble size, but large blocks are common. Note the orientation of twigs and branches, and that many are caught against the tree trunks. Most of the larger trees show old scars where small bonfires from the accumulated twigs burned the tree at the time of a woods-floor fire.

The Grassy Knob chert commonly is exposed as a low, discontinuous bench around the hillsides. Large blocks are found on the slopes and in the steep, stream valleys below. The rock is colorfully stained—yellow to red—and spongelike from the solution of limestone.

Several papers and theses at Southern Illinois University-Carbondale have been written about the vegetation and the relationships between the plant associations and the physical environment. A number of hill prairies survive along the Mississippi Bluffs and smaller grassy openings persist on some of the other high ridges. They seem to be losing to the trees, perhaps because of fewer fires and a period of wet years.

The Shawnee Hills topography is dominated by ridged uplands with many cliffs and deeply dissected valleys. The main cuesta ridge of the northern Shawnee Hills extends from the Mississippi River to the Ohio. The steep south-facing escarpment is nearly in the middle of the division (see Fig. *87*). It is compound in the sense that the massive sandstones form successive ridges separated by benches underlain by shale. From the high east-west ridge many parallel northward trending ridges, separated by deep valleys, slope gradually

87. Looking west from High Knob, Hardin County, along the Pennsylvanian sandstone cuesta. Garden of the Gods is on the escarpment crest in the middle ground.

88. Shawnee Hills cuesta ridge looking southwest from above Eagle Creek Valley in Gallatin County. Pounds Hollow Lake is at left middle ground.

toward the plains (Fig. *88*). The escarpment itself is much dissected on the south by deep valleys such as Lick Creek and Bay Creek, or by deep ravines like the head of Buck Branch (see Fig. *89*). No river valley extends north-south across the escarpment.

The ridges of the southern part of the Shawnee Hills (Fig. *90*) are in themselves small cuestas with their steep slopes facing southwestward and gentler backslopes toward the northeast. Each ridge has a cap of resistant sandstone, commonly cliff forming. Silty loess generally mantles the rock but no glacial till is present south of the Pennsylvanian escarpment. Remnants of an ancient soil lie between the rock and the loess. In road cuts across the highest part of the ridges at the contact between loess and rock, look for well-rounded quartz pebbles with red vein markings. The bedrock has no pebbles; the scattered pebbles must be remnants of some ancient gravel deposit.

The cuesta ridges in the southern Shawnee Hills are intersected by watergaps (Fig. *5*). Streams which have been flowing on weak rocks parallel to the ridges suddenly turn and cut through them in narrow "gorges." Their origin probably dates back to a time when coastal plain deposits buried the ridges altogether. Later, when erosion began again, streams deepened their valleys into the loose coastal plain sediments until they reached the resistant Paleozoic rocks. Once established, the river could not leave its valley. It remained in its course eroding through the resistant rock. Subsequently, during the past 40 million years or so, the coastal plain deposits were eroded away and the present topography was etched

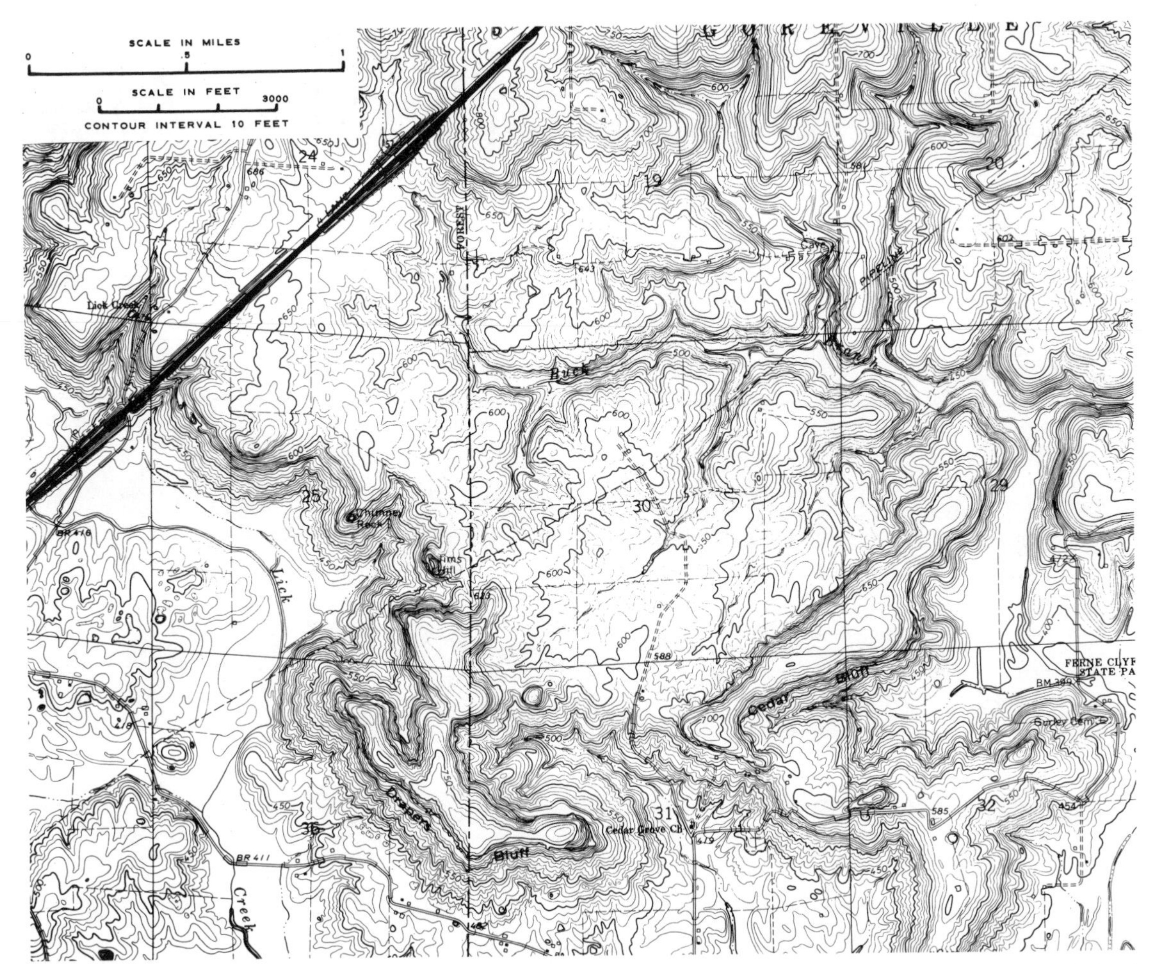

89. Lick Creek topographic map showing part of the Pennsylvanian cuesta that extends from the Mississippi Valley at Horse Shoe Bluff, Jackson County, to Battery Rock, Hardin County. Draper's Bluff is a massive pebbly sandstone bluff. The same Battery Rock sandstone forms cliffs where the valleys have eroded into the hill lands. Deep road cuts and fills of I-57 extend across the northwest corner.

90. Sandstone scarp face in Pope County. The ponds are man-made.

into the rock. Resistant sandstones eroded more slowly and became the ridges, while the more easily degraded outcrop bands of shale and limestone became the valleys.

Bedrock Hills of the Plains

McLeansboro 1:62,500

Hilly lands mark the Mt. Vernon Hill Country. The hills may have steep slopes, and the relief from valley bottom to hilltop is two hundred feet or more in parts of Hamilton County (see Fig. *91).* These are bedrock hills mantled with glacial till and wind-deposited silts (Fig. *92*). Resistant sandstone layers cap the hills and crop out along stream valleys and in low bluffs. These hilly areas display many of the attractive features which characterize the Shawnee Hills, though on a smaller scale.

Look for exposed rocks. Sandstones will be easiest to find but thin limestone layers and shales may be found where erosion is active.

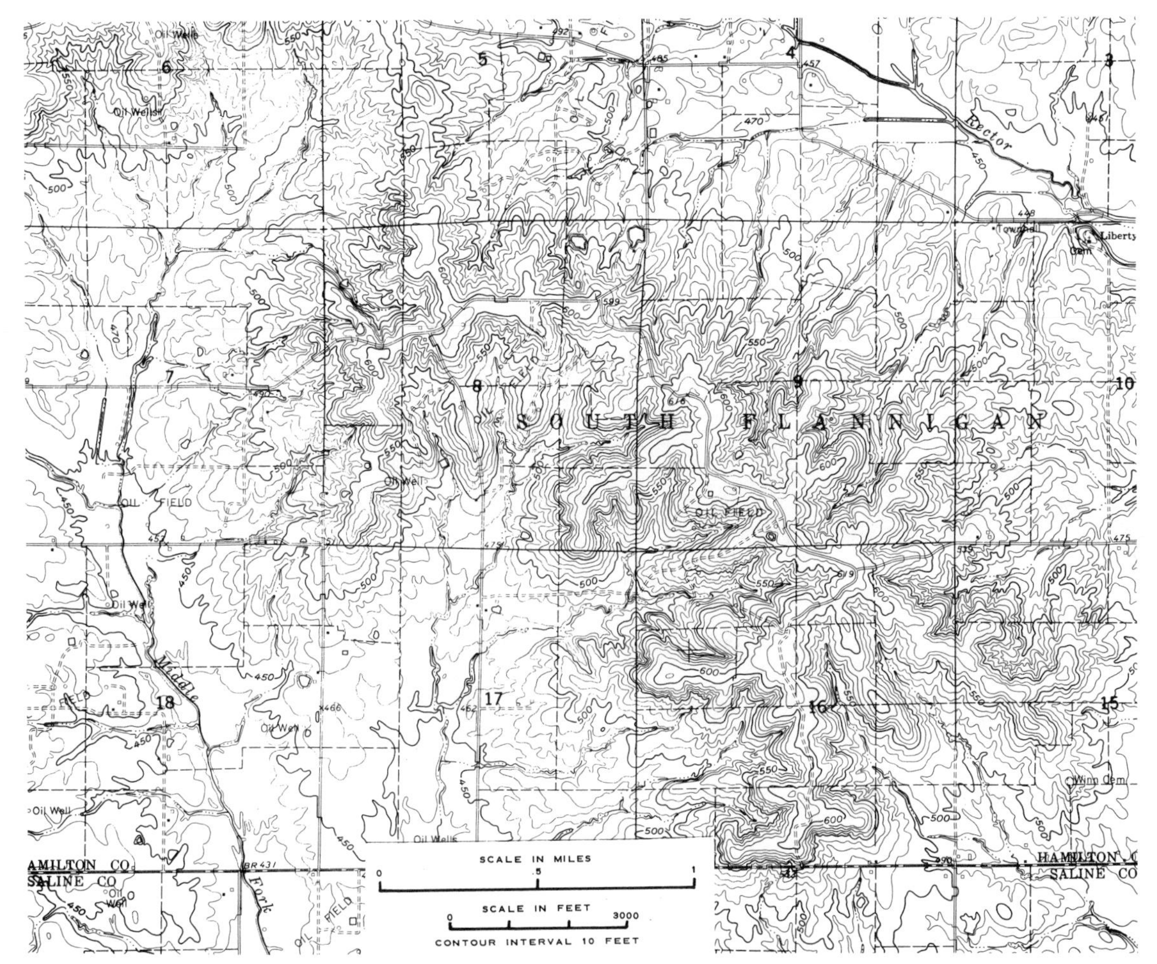

91. Hilly lands of the Mt. Vernon Hill Country shown on the Akin Quadrangle in Hamilton County. These are bedrock hills veneered by loess and glacial till.

92. Mt. Vernon Hill Country southeast of Carmi, White County, showing cultivated rolling hills and wooded stream valleys. Erosion is severe in some places. These are bedrock hills.

Cliffs and Bluffs

Lick Creek 1:24,000
Harrisburg 1:62,500

Cliffs and bluffs are outstanding attractions of Southern Illinois. They crown many a high ridge overlooking forest and pasture below, and present a view of distant countryside. They line the valley of hundreds of creeks and enclose their tributary coves and hollows.

The Shawnee Hills Division (Fig. *1*), which contains them, stretches from the Mississippi to the Ohio river valleys, from Randolph County to Gallatin and Hardin counties. Within the pro-

93. Draper's Bluff at the Union-Johnson County line, a massive sandstone cliff forms part of the Pennsylvanian cuesta. Photograph shows horizontal bedded layers and widely spaced vertical joints.

vince numerous parks and recreation areas provide ready access by road and well-marked trails to the overlooks. Some especially beautiful and fragile cliff-enclosed valleys are set aside as nature preserves. Some cliffs are directly accessible by good roads, while others are little disturbed and rarely visited.

The rocks composing the cliffs are all sedimentary and mostly of massive sandstone. Limestone also forms spectacular cliffs along the Mississippi and Ohio river valleys, notably at Alton, Prairie du Rocher, Pine Hills, Cave in Rock, but does not form cliffs in the hill country. Shales are too readily weathered and eroded to form cliffs. These rocks tend to alternate and the units are of variable thickness. The highest cliffs (100 feet or more) follow the southern boundary of the Pennsylvanian System and form the cuesta escarpment. To the south, the Chester Series consists of alternating formations of sandstone-shale and limestone-shale. The sandstones form many beautiful cliff sites. Across the south of the Shawnee Hills Division, especially in Union and Hardin counties, limestone is the dominant bedrock but cliffs are low and discontinuous.

It is not just that massive sandstones form cliffs. The hilly nature of the Shawnee Hills is due to the fact that in humid climates they resist the weathering and erosion processes more successfully than shale and limestone. The massive, well-cemented sandstones are the ridge formers, the shale and limestones are the locus of subdued relief and lowlands.

94. Buck Knob in eastern Union County, an isolated hill or knob capped by blocks of resistant sandstone. In the distance at the right is the Pennsylvanian Escarpment from which the knob was separated by erosion of Lick Creek valley.

The cross section (Fig. *4*) represents a north-south slice across the Shawnee Hills. Alternation of resistant and weak rock units cause alternating ridges and lowlands. Many of the ridges have cliffs along them. Thus, cliffs are not isolated features. Draper's Bluff, Garden of the Gods, and Battery Rock are outlooks formed from the same sandstone formation coming to the surface at the Pennsylvanian escarpment. Similar lines of cliffs mark other resistant sandstone outcrops.

The Lick Creek map (Fig. *89*) shows how winding and irregular the cliffs are. Cliffs of the Battery Rock sandstone occupy a band a mile or more in width. The many valleys which are eroded into the escarpment cause the irregularity. At the headlands great promontories such as Draper's Bluff (Fig. *93*) jut outward. Where erosion has cut through the neck of the promontory an isolated hill is cut off. The great sandstone blocks at the crest look like the ruins of some ancient castle. Buck Knob near Lick Creek, Union County, is a good example (Fig. *94*).

In places the cliffs grade into steep slopes with a succession of rock ledges. This is because the massive sandstone grades laterally into thin-bedded sandstones and shales which will not maintain a cliff face. Cliffs form in hilly lands where massive sandstone layers resist

95. Sandstone cliff at Giant City State Park, Jackson County. A stream channel formerly flowed against the cliff in the background. Note the horizontal bedding and the numerous weathered pockets. The top of the cliff has only a thin soil cover and is a very dry environment.

the processes of weathering and erosion more successfully than the more shaley rocks lying above and below them. Generally there is a rounded soil-covered ridge above, then a rocky ledge and precipitous bare rock face, followed by a steep slope below. The cliff is maintained because the weak rock below is constantly eroding and because the debris that falls from the cliff face does not accumulate against it at the base. Erosion on the free face itself proceeds as rapidly as weathering releases particles; no soil accumulates. Some cliffs extend down to the level of the valley floor where streams can undercut them and directly remove fallen debris (Fig. *95*).

At the top of most cliffs bare rock forms a zone a few feet to many yards wide (Fig. *96*). There is clear evidence of active erosion of the stony weathered mantle and the overlaying loess. Such erosion has been greatly hastened by man's use of the land permitting more rapid runoff, trampling of the undergrowth and grazing by animals, and cutting of the woodland cover. There is much of interest on these ledges where plants struggle to find soil and moisture, where water seeps out beneath the soil and rainwater forms torrents and waterfalls over the bare rock. Here, too, the big cracks or joints, are visible. In massive rocks they are widely but regularly spaced, generally in sets

96. Bare sandstone above a cliff. The loess and soil have washed away, but cracks and depressions retain some earth and water. Here plants cling to life through the hot summers, and interesting dry climate forms are common.

at right angles to one another. Most often one set is at right angles to the cliff face, the other parallel to it. If there is a means to climb down the cliffs, the joints are the most likely passageways.

Viewed from below at a distance of one hundred yards the vertical joints and horizontal bedding planes give form and character to the cliff (Fig. 97). The horizontal layers mark successive depositional events. Many of the horizontal layers are segmented by sloping "cross-beds" formed by the growth of sandbars or the filling of scour channels when the sandstone was deposited 300 million years ago. The vertical joints are fractures formed in the distant past as the great pile of rocks was strained by the gradual upward bending of the Illinois Basin. The fractures would be invisible when buried beneath the weight of hundreds of feet of rock, but they gradually widen when erosion releases the pressure. Near the surface, water, then roots, penetrate the fracture and gravity tugs the block downhill. The pattern of joints is clearly visible at Bell Smith Springs or Giant City where separation of great blocks has widened the joints into crevasses and finally "streets." Great blocks have slid and tumbled away from most cliff faces (Fig. 98).

97. The vertical joints result from fractures caused by stresses during the warping of the Illinois Basin 200 million years ago. The joints are zones of weakness along which weathering and erosion take place. Horizontal lines are the bedded layers of sandstone. Protruding zones are more densely cemented with brown iron oxide.

98. Large blocks of sandstone in creek bed near Makanda, Jackson County. The steep slope and a surface of clay and silt—slick when wet—allowed gravity to draw the blocks downhill from the cliffs above.

Bluff Shelters

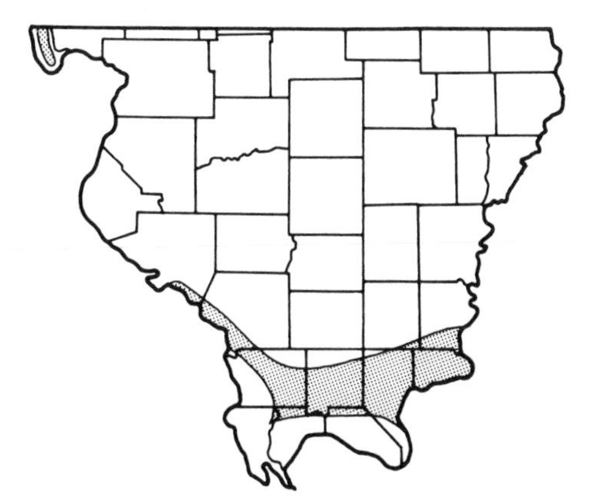

Thousands of bluff shelters may be found in the Shawnee Hills Division associated with the high cliffs. Some are readily accessible along roads or in parks, but most are reached on foot (Fig. *99*). A walk which includes cliffed hills will almost certainly encounter a shelter. A bluff shelter holds a special fascination and a wondering thought about people back through the ages who sought protection from a storm or perhaps made camp while hunting or gathering nuts and berries.

Shelters range from small hollows to great half-domes. Big shelters, 100 yards long, 75 feet high, and 75 feet deep, are not uncommon. A half arch forms as blocks and pieces of the roof collapse (Fig. *99*). Collapse is initiated above a zone where the sandstone is more subject to weathering and disintegration. The sandstone bedding is thinner or contains shaly partings, or the grains of sand are less well cemented. This zone of weakness may cause a series of shelters at the same horizon on a valley side or perhaps on opposite valley walls.

A few shelters of this type are found in limestone as, for example, the famous Modoc archeological site along the Mississippi Valley bluff near Prairie du Rocher, Randolph County. This is an overhang—not a true cave. True caves are dissolved from the rock and have more extensive and intricate passageways. Some bluff shelters are marked as caves on maps and are well known by that designation, such as Saltpeter Cave, south of Murphysboro, Jackson County.

Topographically most bluff shelters are on a hillside in the base of a cliff. They may look out far across the valley or the lowland below, or they may be within an enclosed valley. Where the massive sandstone bluffs approach the valley floor, shelters may be flooded in high water (Fig. *100*). At such positions stream erosion hastens the roof collapse.

Examination of the rock reveals the features characteristic of massive sandstones: widely-spaced joints, cross-bedding, ripple marks, concentrations of dark iron oxide cement along the bedding planes or in irregular festoons, peculiar boxwork and cupped surfaces. (These are described under their own headings in sec. Primary Sedimentary Features.) Sedimentary features of the underside of sandstone beds may be especially well exposed in the ceiling of the overhang.

The floor of a shelter consists of rocky layers partially covered by sand and silt. The loose sand comes from disintegration of the walls and roof of the shelter itself. Grains are separated by freezing and thawing of water in the rock pores or by the decomposing of cementing material. In some places the sandstone is very friable and fragments crumble in your hand. The silt comes from the loess above, some washed down the cliff face and directly into the shelter, some washed into the shelter by surface rivulets flowing along the base of the cliff. Blocks and angular fragments are scattered about; some are freshly broken, matching fresh breaks in the ceiling. Roof collapse still goes on. The broken blocks gradually disintegrate and the sand grains are washed away.

Channels of surface water mark part of the floor in many shelters, even those high on a hillside. In some places the source is a small spring or seep at the back of the shelter, usually where a fracture extends upward. During wet seasons there is flow, but not in dry seasons. More often water comes from rain and surface runoff (Fig. *101*). At times of heavy rain a sheet of water clings to the cliff face and is drawn into the shelter. A line of drips can be identified by splash craters in the floor.

At the edge of many bluff shelters a low ridge of silt and sandstone blocks forms a protective wall between shelter and the steep hillslope below. Most of the debris came from above, loosened by animals, plant roots, and weathering processes, it tumbled over the cliff but lodged in the mouth of the shelter.

To appreciate the bluff shelter one should find a way to the cliff top above it. The slope continues steeply upward, perhaps marked by gullies and exposed rock. In some places small stream channels end abruptly at the cliff's edge. During winter and spring a considerable waterfall plunges over and forms a plunge pool below. The larger falls have caused some cliff retreat and the plunge pool may once have served as a convenient water source for people who used the bluff shelter.

The particular atmosphere and microclimate of a shelter bluff depends on its outlook and exposure (Fig. *102*). Those which face southward are rather dry and receive the winter's sun. They are also dry and hot in summer.

99. Bluff shelter at Bell Smith Springs, Pope County. The stone wall is modern. The shape of the ceiling is typical. Note the fallen blocks in the middle distance and, in the far distance, the large, tilted blocks that have fallen away from the cliff.

North-facing shelters are moist and cool, especially if located in an enclosed valley. The sun never reaches them, so temperatures remain moderate and lag behind the season. Evaporation is low. Spring and early summer are especially damp periods because the rocks remain cool. Warm, moist air circulating through the shelter causes condensation on the walls and floor. Lovely and delicate flowers and ferns may be found in such a special ecologic niche. Here too, the environment is favorable to algae, liverworts, and moss.

100. Cliff at Ferne Clyffe State Park, Johnson County, showing reentrant, at least partly caused by stream action.

Joint Blocks

Great blocks of sandstone standing near a cliff or resting far down the slope are striking features which raise considerable curiosity. Most blocks are bounded by rather regular and straight sides (joints or rock fractures), though some are curved. Generally their shape clearly fits the adjacent wall of the cliff (Fig. *103*). What caused the separation?

Where a stream flows close to the base of the cliff as at Ferne Clyffe, Johnson County (Fig. *104* and *105*), it is easy to surmise that flowing water eroded weaker rock at the base. Finally the massive rock, lacking support, broke away along the joint faces. Most cliffs and blocks are high above the valley floors. The separating process is not so clear, but the same principles hold. Weaker rock beneath the massive units is eroded by drip and fall of water from the cliff top and by disintegration and decomposition. The pull of gravity causes slow downhill creep of loose material on the steep slopes. Eventually lack of support causes a fall (Fig. *105*).

More curious and less easily explained are the "streets" as seen at Giant City, Jackson County, and many other places in the Shawnee Hills (Fig. *106*). Here the blocks appear to have moved apart horizontally, maintaining vertical walls. The bedding planes between cliff and block match with little vertical displacement. Possibly the movement dates back to Wisconsinan time, 12,000-plus years ago, when the ground froze to greater depths and snow and ice accumulated between block and cliff. Freeze and partial thaw and refreezing would exert great expansion stresses and at the time of spring thaw the wetted clayey rock underlying the sandstone units would provide a surface with low friction. Observations over a period of twenty-five years have not revealed evidence of recent movement of these blocks.

The surface of many of the blocks exhibit curious holes and pockets surrounded by resistant brown bands and involutions (Fig. *107*). Some joint faces have a nearly continuous dark brown "crust" penetrated here and there by pockets. The brown material is iron oxide "cement" which fills the pores between sand

101. Bluff shelter at Kerr Bluff, Union County. Note the slope of the ground is into the shelter. Bare rocks cover the floor because running water has removed the fine sediment.

102. A small, southwest-facing shelter bluff at Dixon Springs State Park, Pope County. In spring it is very moist but becomes dry in summer. Note cross-bedding in the sandstone.

104. Great blocks separated from the cliff along the joint faces. These are in Ferne Clyffe State Park, Johnson County. A stream channel is in the foreground.

103. Ferne Clyffe State Park, Johnson County, has many fascinating examples of sandstone cliff-related features. A narrow "street" has been formed by the separation of a large block. The joint opening in the upper part of the photograph will someday break, causing the fall of another large block.

105. Large block broken from cliff along a joint in the massive sandstone. The honeycombing of the joint surface was made by resistant concentrations of iron oxide surrounding pockets of poorly cemented rock.

106. A "street" caused by outward movement of sandstone block from the cliff face. Smaller joint blocks have subsequently broken off from the cliff. Bell Smith Springs, Pope County.

grains. The iron oxide cement is present throughout the rock but tends to be concentrated along joint faces and bedding planes. Such concentrations may be so dense that it is hard to break with a heavy hammer; when finally broken the fracture may extend through, rather than around, the quartz sand grains. The pockets are places from which groundwater removed the cement. Thus, when exposed to weathering, the grains are washed or fall away. Note that many holes are now lined with well-cemented rock.

Natural Bridges

The best known of the natural bridges are near Pomona, Jackson County, and at Bell Smith Springs (Fig. *108*). Both are sandstone. Numerous small bridges between the great separated blocks and adjacent cliffs can be found. Natural bridges in the limestone terrane of Southern Illinois are not known to the authors, but they may exist.

The length of the span of the two large bridges is their most startling feature (Fig. *109*). The span, in essence, represents one great joint block. The sides are even and straight and no cross-joint fracture is visible to form a plane of weakness. Both bridges are close to massive cliffs and upslope from the main stream. Nevertheless a rocky stream channel passes under each bridge (Fig. *110*). Erosion is clearly deepening this channel and slope wash is gradually undermining the end supports. The arched roof at Bell Smith Springs suggests that it may once have been a bluff shelter. The most probable origin of the bridges is suggested by the stream channel. Long ago the front of the span formed the cliff face. Water from uphill percolated down a parallel joint and found its way out along a bedding plane. Such a seep or trickle of water can be seen in springtime at the back of many bluff shelters. Eventually an underground channelway opened up, and weathering, mass-wasting, and erosion carried material down the joint and under the incipient bridge. The massive size of the joint block made it resistant to collapse. The spans will remain until the end supports give way.

107. Sandstone block showing irregular holes and pockets. Iron oxide tightly cements the sand grains along the joint face. Where cementing is less dense, weathering has loosened the sand grains and they have washed away.

108. Part of the Stonefort topographic quadrangle showing the site of Natural Bridge at Bell Smith Springs and Burden Falls in the north.

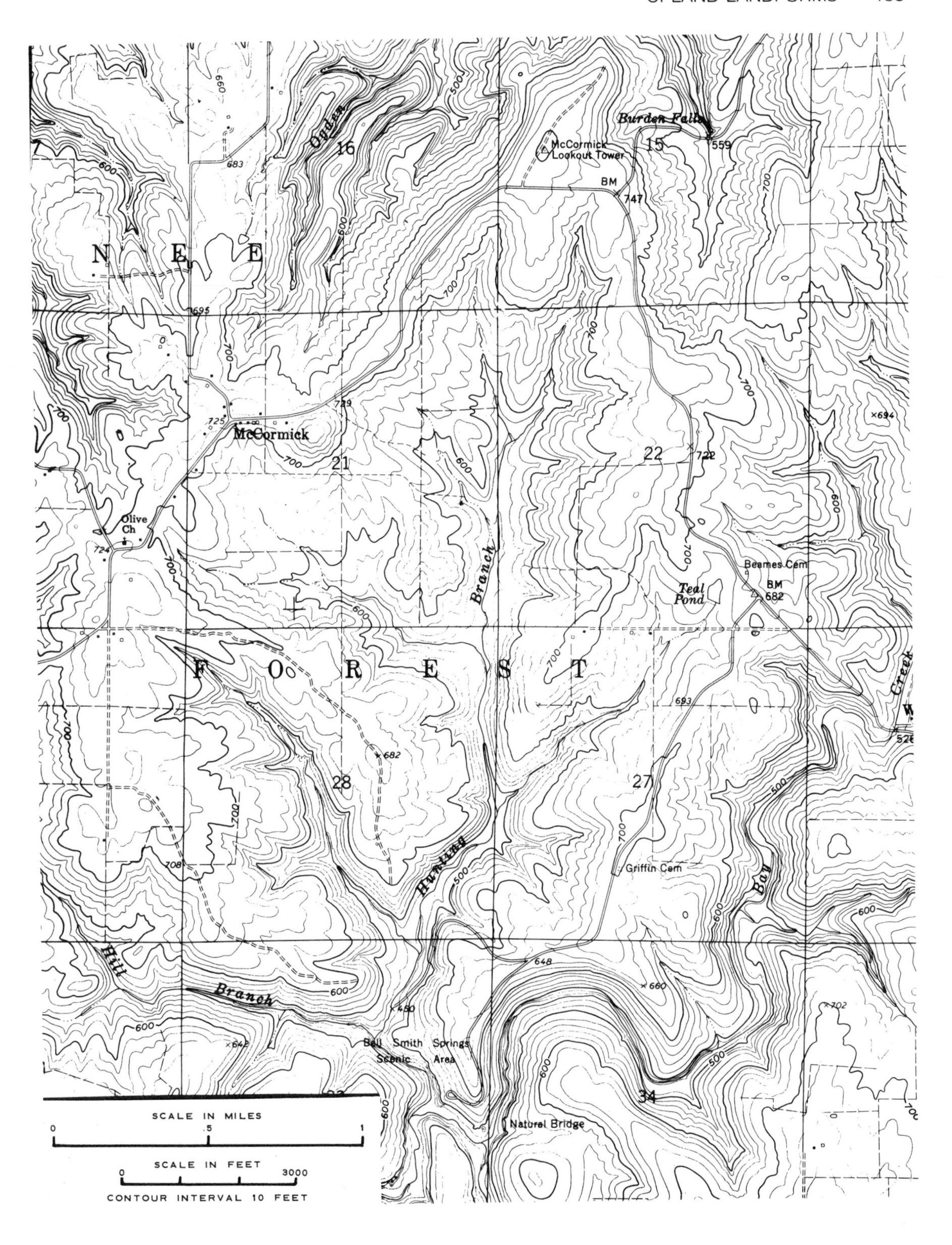
Ogden
16
McCormick Lookout Tower
Burden Falls
15
559
BM
747
N E E
695
McCormick
725
729
21
22
722
Olive Ch
724
Beames Cem
BM
682
Teal Pond
Branch
F O R E S T
Creek
693
682
28
27
Hunting
Griffin Cem
Bay
Hill
Branch
648
660
450
642
702
Bell Smith Springs Scenic Area
34
Natural Bridge
SCALE IN MILES
0
.5
1
SCALE IN FEET
0
3000
CONTOUR INTERVAL 10 FEET

109. Natural Bridge at Pomona, Jackson County. Note its straight sides and the stream channel beneath the span.

110. Looking downslope under Natural Bridge. Supports show arching form and remains of the shelter at far end. A seep of water along the joint gradually enlarged so as to carry eroding storm runoff beneath the great joint block.

Limestone Terrane

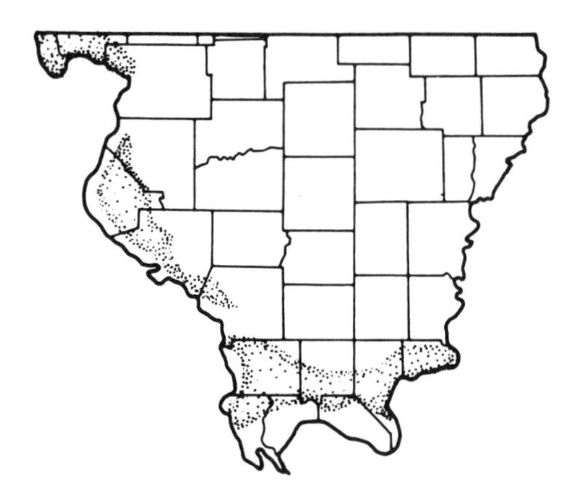

Waterloo 1:62,500
Renault 1:24,000
Cave in Rock 1:24,000
Anna 1:24,000

Limestone regions have distinctive topographic features which make them particularly attractive for exploration. Such areas are called "karst" after the Karst region of Yugoslavia and Italy. They are most notable for caves and springs, but collapse sinks, disappearing creeks, cliffs and trenched valleys, gravel-filled stream channels, and interesting residual soils add adventure and excitement to a day's outing. Large quarries provide crushed rock and fossil and mineral collections.

The accompanying map from the Waterloo area, Figure *111*, and the aerial photograph from near Renault, Figure *112*, both in Monroe County, illustrate the features typical of our limestone terrane. The upland is marked by round and irregular depressions and an absence of stream channels (Fig. *113*). Some surface streams flow a short distance, then disappear into a sinkhole. In Monroe County a regular underground drainage system of parallel caverns, each with its collecting tributary passages, gathers seepage water. The streams reach the surface at springs, some below the level of the master streams. In Union and Hardin counties most of these springs provide entrances, above water level, to the caves.

A karsted topography develops in humid regions where there is a thick unit of dense limestone bedrock (Fig. *114*). Well-developed bedding planes (Fig. *115*) and a vertical-joint system will provide passageways for groundwater movement. Lastly it is thought that a deep master valley will promote rapid circulation of groundwater. Dense rock with a pronounced fracture system promotes concentration and solution in localized zones while providing strength to support open passages.

Downward seepage of water along the fractures permits solution of the rock and gradual integration of and underground flow of the water. Vertical passages or sinks carry surface water downward to the water table where myriads of openings hold the water until it seeps into larger passageways which eventually carry the water to a surface spring.

In Southern Illinois the limestone bedrock is covered first by a residual soil consisting mainly of clay and siliceous and cherty material. This soil cover fills many of the dissolved openings in the rock and slows or inhibits infiltration from the surface. The additional cover of glacial till north of the Shawnee hills and loess over all the limestone terrane moderates the rate of infiltration of water and subdues the severity of karst topography.

111. Topographic map of Waterloo Quadrangle in Monroe County. The upland is pocked by myriads of depressions and sinkholes. Fountain Creek has a good flow of water, but the valley at south center is discontinuous and loses flow underground. There is a large spring outlet at Camp Vanderventer.

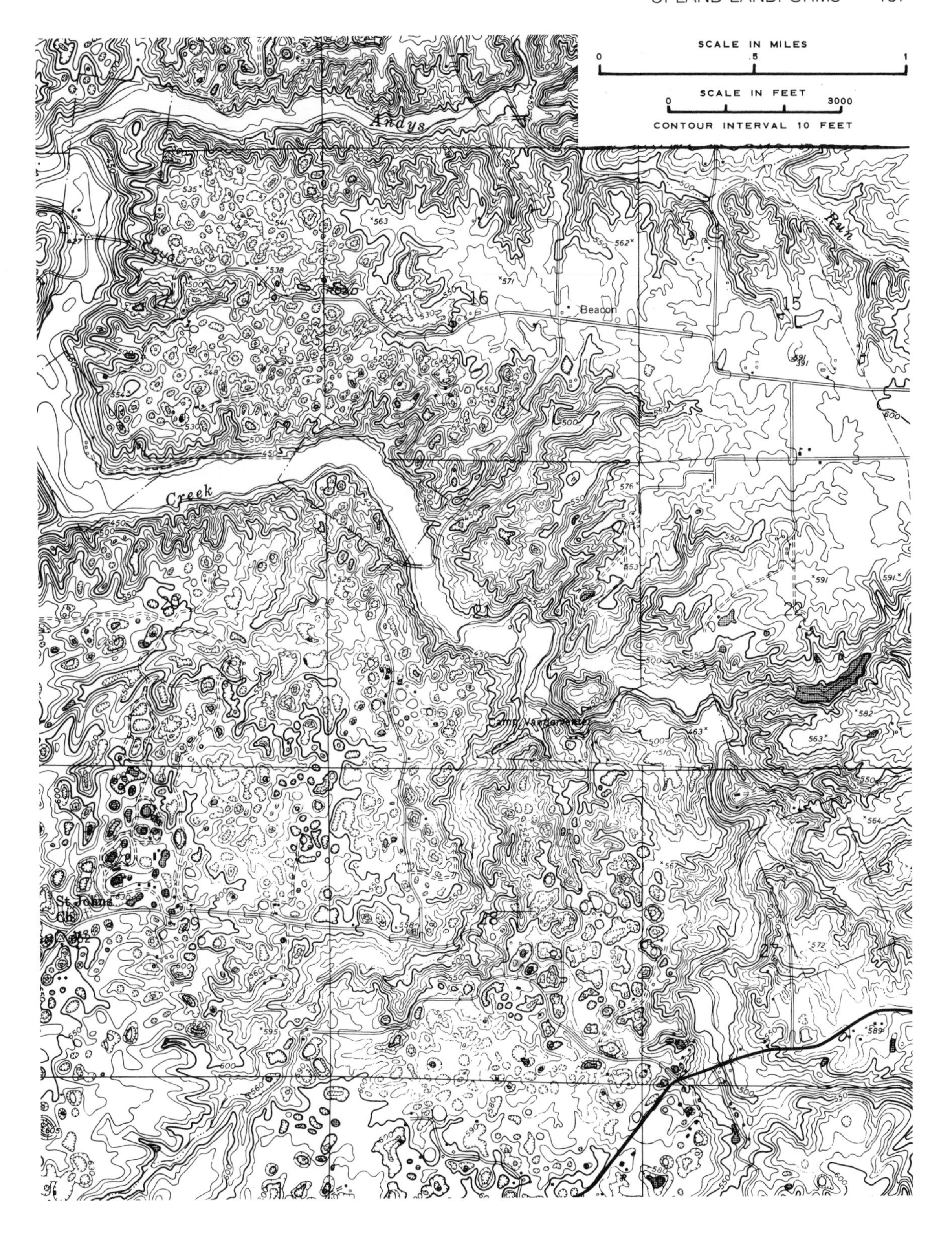
SCALE IN MILES
0
.5
1
SCALE IN FEET
0
3000
CONTOUR INTERVAL 10 FEET
Andys
Run
Creek
Road
Beacon
Camp Vandeventer
St Johns Ch
16
15
17
20
21
22
28
27
29

112. Sinkhole plain near Renault, Monroe County, looks like a bomb-cratered battleground. The wooded valley in the background is at the edge of the limestone bedrock. Note gullies and waterways leading to sinkholes and small swamps or lakes. There is no integrated surface-drainage net.

113. Rolling karst land near Renault, Monroe County. The pond is a sinkhole, plugged up by organic debris and soil wash. It may suddenly drain away underground. No stream is present.

114. Close-up of dense limestone showing solution passages along horizontal and vertical fractures. The intricate anastomoses above the large cavity mark a zone frequently flooded by moving underground water. Hutchins Creek, Union County.

115. Thin-bedded but dense limestone at Krueger Cave, Monroe County. The passageway developed along a bedding plane is also controlled by vertical joints.

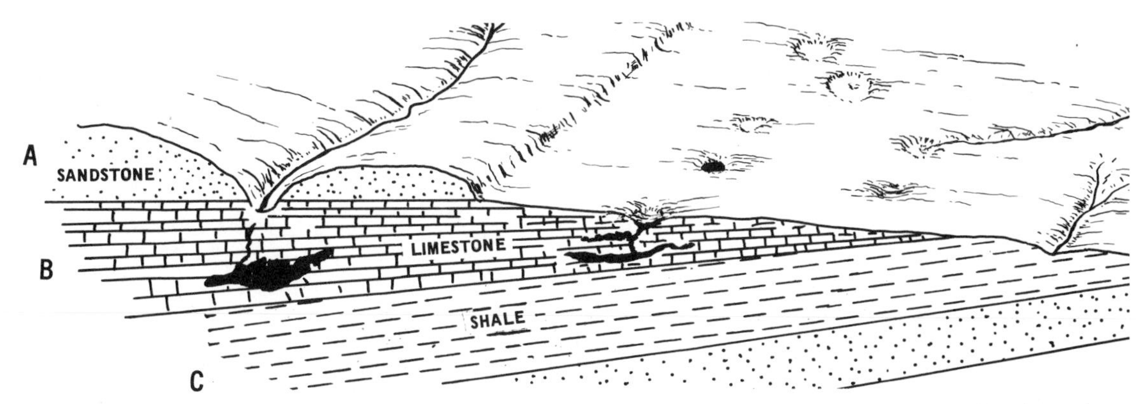

116. Diagram representing a limestone terrane with sinkholes, sinking streams, caverns, and springs along the river valley. Normal streams flow on shale and sandstone.

The limestone sedimentary rock layers, which are subject to solution and a karsted topography, are overlain and underlain by insoluble rocks on which normal drainage patterns develop. The most obvious indicator of karst is the sinkholes. On the diagram, Figure *116*, some sinkholes might form where unit A is thin because of collapse into unit B. However, unit C would be devoid of sinkholes. Streams flowing from C to B might disappear into underground passages.

Some larger valleys have eroded through the limestone or flow across less soluble rock. These are characterized by limestone cliffs and rocky stream channels. Fountain Creek in Monroe County is a good example (Fig. *111*). Streamflow sometimes disappears beneath the channel and springs are present along its course. Smaller valleys may end in a blind sinkhole as in section 29, Figure *111*. Accessibility to the cavern below depends mainly on the state of accumulation of logs, organic debris, and soil material in the sinkhole.

SUGGESTED READING

Lamar, J. E. 1959. *Limestone Resources of Extreme Southern Illinois*. Urbana: Illinois State Geological Survey, Reports of Investigations no. 211. 49 pp.

Sweeting, M. M. 1973. *Karst Landforms*. New York: Columbia Univ. Press. 362 pp.

Caves

A cave has been defined as a hollow place underground. It might be barely large enough to hold a person or be very large with an intricate system of passageways. Most Southern Illinois caves are in limestone, some rather extensive, but several sandstone caves fit the above definition. The topographic maps and popular usage identify many of the bluff shelters as "caves," but these are considered a distinctly different feature.

Caves in limestone (Figs. *117* and *118*) result from the dissolving of the rock along fracture zones and bedding planes. They form near the earth's surface where moving groundwater carries dissolved carbon dioxide and organic

117. Cave passage near entrance to Illinois Caverns south of Waterloo, Monroe County. Note rounded passage walls at the right. On the left there has been collapse of the shelf and also the ceiling. The slope in shadow is a fill of silt deposited at high water by the underground stream.

acids. Percolation of water is most rapid above the water table (see diagram Fig. *119*) and near the top of the zone of saturation where the groundwater is drawn laterally by gravity toward the nearest valley. Some solution certainly takes place at greater depth, but the water becomes neutralized and less chemically active. The water table, the top of the zone where all the openings in the rock are filled by water, fluctuates up and down according to the abundance of rainfall. Many caves are located at this zone. At times of rapid runoff some parts of caves become filled with water because of small outlets or restricted passageways. Grass and twigs on high ledges attest to recent flooding. Caves exist below the water table, especially in Florida; in Southern Illinois, there may be such caves at Pine Hills beneath the springs.

The largest cave entrances are near the valley floor and may have a stream or spring issuing from them (Fig. *120*). Generally such an entrance leads to extensive passageways, although the passage in Cave In Rock slopes down below the Ohio river level and hence is filled with sediment. Most entrances are through sinkholes and hillside openings into the middle of the passage system. In Monroe County most entrances require a rope to help manipulate the ninety feet or so down to, and later back up from, the floor of the cave.

Exploration of a cave is always a challenging

experience. It is like entering a different world. Even at the entrance the atmosphere is different. In summer the air feels cool, in winter, warm, because the interior temperature remains nearly constant at the average annual temperature of the area—for Southern Illinois 56° to 59° F. The air is moist, and light fades rapidly away from the entrance until around the first corner—total darkness. Eerie shadows are cast by the explorer's artificial light. Some caves have high ceilings and large passages, so there is a feeling of spaciousness (Fig. *118*), but most passages are narrow and the ceilings are low. Here one must crouch, even crawl. The floor of most caves is the bed of a stream with gravel bars. For a time one may avoid wet feet but eventually a pool extends from wall to wall and continuation does mean getting wet. Then the ceiling slopes down and water or mud covers the floor; that is the moment of decision. Once wet, you can relax! Still, a large room or a dry upper level brings a feeling of relief. As long as one keeps moving the moisture and temperature are not uncomfortable, but discomfort sets in as the body loses heat after resting for a time or while searching slowly and taking notes and measurements.

118. Main passage of Layoff Cave near Rosiclare, Hardin County. The passage is about fifteen feet high.

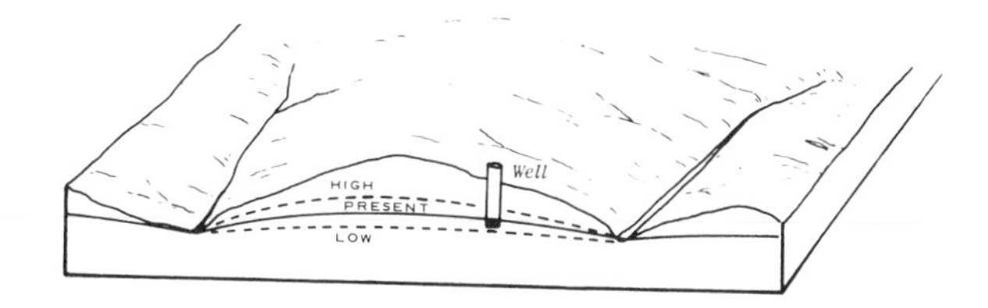

119. The Fluctuating Water Table

There are many interesting cave features. Interesting biological organisms (salamanders, cave crickets, and spiders, algae and fungi, racoon tracks and droppings, colorless shrimp and small fish, and so forth) can be seen, but the focus here should be on the wide variety of erosional and depositional phenomena. More thorough descriptions will be found in some of the references. The cave itself is the result of erosion by the dissolving action of moving groundwater. Evidence of continuing solution is seen in smooth and fluted walls and the rock floor (Fig. *121*). Thin blades of rock may be razor sharp. Look at the smooth surfaces where water flows against it and the solution-rill channels down the walls. Sections of fossils are often visible. Slightly siliceous fossils stand in relief. The upper parts of cave walls are much rougher, since water does not wash across them.

Deposits by cave streams are in the form of gravel and sand or fine muddy silts. Pebbles are of chert released from the limestone walls. Deposits of calcium carbonate ("formations")

120. Keith Cave and Spring north of Anna, Union County. A good stream flows even in dry seasons.

are precipitated by partial evaporation of the film or drip of water from walls and roof. Stalactites hang from the ceiling in "soda straws" or massive pendants, and stalagmites rise in mounds from the floor and crusts of flowstones may cover the walls and sediments. Rimstone pools and, rarely, "cave pearls" in drip pools are also known (Figs. *122* and *123*).

The pattern of passageways in a cave or cave system is very interesting. The larger caves tend to be linear, but frequent near-right-angle turns are common. Orientation is determined by the joint system. Note the many narrow solution openings parallel to one another at successive joints on each side of the main passage. Many caves (Frieze, Cave Hill) have parallel passages with cross connections giving a network pattern. Such caves were formed under conditions of saturation or very slow water movement, not by the eroding action of underground streams.

In many caves today underground streams are active eroding agents with many of the characteristics of surface streams. They carry much sediment and erode and scour the sidewalls. Widening brings about roof-fall (Fig. *117*). Periodically along the passages there are high domes or side rooms leading to high domes (Fig. *121*). The walls have solution grooves (lapies) and perhaps a film of water. Water commonly drips from a jumble of large blocks of rock, tree trunks, and finer debris forming the roof. The ground surface is usually close above and sometimes passages to the

121. High dome above the cave passage in Illinois Caverns. The walls are scalloped and fluted by solution caused by the film of water that descends from the surface. Part of a small tree carried down the overlaying sinkhole is visible in the dome.

122. Active flowstone dome in Layoff Cave near Rosiclare, Hardin County. The attractive small "organ-pipe" formations and rimstone pools have not been defaced. Although visited by hundreds of people every year there is little vandalism.

surface exist. Occasionally such a roof collapses.

During the Pleistocene glacial epoch most cave passages became filled with gravel, sands, and silts. Lilly Cave is still mostly filled. Present-day cave streams are eroding this detrital fill and slowly reexposing the ancient rock passageways. In some caves, Rich's for example, new and smaller crawlways bypass the original passageways still filled with detrital sediment. Ava Cave displays stream-eroded troughs intersecting one another in the solid rock. Gravel deposit remnants are tightly cemented into some troughs.

Limestone caves yield valuable records of the past. Not only do they record successive intervals of erosion and deposition, but the deposits occasionally yield fossils which tell something of life of the past. Caves are dynamic features which respond to changes in the environment. Today they are actively enlarging, both through solution and mechanical removal of Pleistocene deposits.

At least three dome-shaped caves are known in massive sandstones. They are located in the position of a bluff shelter; in fact, the entrance is through such a reentrant at the base of sandstone cliffs. The domes seem to

123. Layoff Cave near Rosiclare, Hardin County. At the right is part of a large flowstone dome, at the left a solution-marked shelf of bedrock. Passageway about 3½ feet high.

have attained their shape through gradual roof collapse, but the initiating conditions are unknown. The walls are damp, as in limestone caves, and joint openings bring water from above into the caves. The direction of water flow and shallow channel erosion is visible on the floor. Large collapse blocks also cover the floor. Some blocks are friable and sand grains easily rub off. Mechanical erosion of this weathered rock by flowing water is the presently active process of cave enlargement.

Things to do. Caving, or spelunking, is both a sport and a scientific venture. It does have dangers and no one should go alone beyond the penetration of natural light. Be sure that the exploring party has adequate lights and reserves: miner's lamps, gasoline lanterns, carbide lights (and reserve carbide), flashlights and batteries, matches, and so forth. An experienced caver should serve as leader. If you do not know someone personally, get in touch with a Geology Department or with one of the grottos listed below.

Cave mapping, using compass and tape, is a challenging and rewarding activity. Map the directions and turns of the passageways, noting if any are now filled with mud or gravel, and sketch cross sections. Show where surface waters—or raccoons—enter or roots penetrate from the surface. Indicate where there are stalactites and if they are active.

Look for fossils in the walls, and note the rock types which have been dissolved to make the passageways. Some layers will have cherty masses in relief; the roof may be sandstone or have shaly layers.

Reference to the books and articles will help one observe the many erosional and depositional features of caves and to interpret their significance.

Grottos:

Little Egypt Student Grotto
S.I.U. Cavers
Southern Illinois University
Carbondale, Illinois 62901

Southeast Missouri Grotto (SEMO)
Southeast Missouri State University
Cape Girardeau, Missouri 63701

SUGGESTED READING

Bretz, J H., and Harris, S. E., Jr. 1961. *Caves of Illinois*. Urbana: Illinois State Geological Survey, Reports of Investigations no. 215. 87 pp.

Folsom, Franklin. 1962. *Exploring American Caves*. New York: Macmillan. 319 pp.

Sloane, H. N., and Guinee, R. H. 1966. *Visiting American Caves*. New York: Crown Publishers. 246 pp.

Cultural Features in the Landscape

Man's activities have a major influence on the appearance of the landscape. The cutting of the forests, plowing of the prairies, and draining of the marshes and swamps profoundly altered the surface aspect of Southern Illinois. We prize the few accounts of the original aspect of the land left to us by early travelers and pioneers. Settling came before photography, and it is only dimly possible to reconstruct the original appearance. The impact of modern machinery and intensive land use brings about ever greater changes.

Our township and range system of land ownership has resulted in squared patterns of land use and road building, especially in the plains division. Drainage ditches extend the squared pattern. Railways and modern highways add diagonals. Although quite apparent from the ground, the aerial view is particularly striking (Fig. *124*).

Man is continually altering the landscape. Many visible features are excavations and piles of waste. Yet the prominent features against the skyline, especially on the plains, are man-made structures: for example, grain elevators, water towers, transmission towers, power lines, and towns and cities themselves. Ruins of numerous abandoned towns and crossroad stores and old family homesteads are clues to past history but now are fast being obscured by nature.

Highways dissect the land. Even in flat areas much earth and rock is moved to provide drainage and stability. Borrow pits are dug in the flat prairies where the roadbed must be raised, as at overpasses (Fig. *125*). Frequently, the pits become ponds. Here can be found soil profiles. In hilly land, deep road cuts expose remarkable sections of rock (see Fig. *126*) and sometimes superb views where the road plunges across what was once cliffs and valleys. In spite of this there is a certain sameness about dual interstate landscaping. The grade is smoothed out, cutting through hilltops and crossing valleys on a high fill. The vegetation is clipped grass and massed shrubbery.

Side roads today are more interesting. Most are graveled and, though graded, they follow

124. Squared fields and roads on the divide between Skillet Fork and Big Muddy basins, Jefferson and Hamilton counties.

the relief of the land and provide more scenic variety (Figs. *90* and *127*).

Railways also extend straight across the land with cut and fill, but their rights-of-way are narrower and less accessible and their engineering somewhat more conservative. The tunnel and deep cuts by the Illinois Central through the crest of the Shawnee Hills north of Vienna are the most striking features.

Abandoned roads in hilly areas are marked by deep trenches eroded into the loess and/or glacial till (Fig. *128*). Many expose bedrock. Natural gravel is a scarce commodity in Southern Illinois, and crushed rock is a product of modern technology. Thus, the old roads had poor surfacing, became deeply rutted and subject to rapid erosion. Loess erodes readily when stripped of vegetation but does have the peculiar property of retaining nearly vertical slopes. The results are visible throughout the hilly areas and are not unattractive. Many old trees cling to the steep slopes although roots are exposed and undercut.

Abandoned railways and intercity trolley lines can be recognized throughout the region. Narrow embankments remain across low places and cuts are found in the valley slopes. Even where no topographic feature remains the trace is often visible in the soils. Aerial photographs show them up especially well.

Pipelines and power lines also leave their trace which is seen more clearly in vegetation

125. A large borrow pit beside I-64 west of Mt. Vernon, Jefferson County. The borrow is now part of the raised roadbed.

126. Long road-cut (a second cut visible in the distance on I-57, a topographic feature in itself. The roads reveal features of the rock rarely visible in natural exposures.

differences than by topographic feature. Deep trenches result in a mixing of mantle or rock material with the soil, so the trace is visible even in cultivated fields (see Fig. 77). A narrow swath through woods and forests identify the trace across the landscape and in aerial photographs. Local distribution systems, such as water and telephone and suburban electric lines, generally follow the road and land-ownership pattern.

Mining and quarrying leave their marks on the landscape (Fig. *129*). We think of the excavations first, but storage piles, waste piles, lagoons filled with fine tailings, water diversion trenches and their spoil, and collapse sinks from underground mines are other prominent features.

The ever increasing demand for mineral resources will bring more rather than less alteration in the future.

Coal beds have a sheet-like shape, very thin but extending over a wide area. At the outcrop (see Fig. *2*), where coal comes to the surface, the overburden of loess and glacial till and ten to a hundred feet of rock can be stripped off by great shovels. The coal is removed and the next band of overburden (now called spoil) is deposited in its place (Fig. *130*). Modern regulations require that the spoil be evened out. Old strip mines (Fig. *131*) were left in long steep-sided ridges containing many ponds. Some spoil areas have gradually become vegetated with shrubs, grass, and trees, but they are very difficult to traverse. Other spoil areas lack limestone to neutralize the acid-forming minerals contained in the black shales. Such areas are barren and release polluted waters to the streams around them. These man-made wilderness areas have become a haven for some kinds of wildlife.

North and east of the strip mines the coal beds slope deeper underground and must be removed by slope or shaft. The principal identifying feature of an underground mine is the tipple and preparation plant. Large piles of coal stand ready for loading (Fig. *132*) after having been separated from the rock material

127. The scenic Rocky Comfort road in Jackson and Union counties, now asphalt and crushed limestone, is graded but still follows the relief of the land.

and the pyrite, FeS_2, mined with it. The fine-grained waste materials are washed out into a holding lagoon. In the old days the coarse materials formed great piles of "gob" (Figs. *133* and *134*). Such waste areas are barren and are sources of continuing pollution. In some places the waste is gradually being removed and in others a cover of soil has been added in hopes of restoring a vegetative cover.

Abandoned underground mines are frequently marked by irregular subsidence of the ground above them. The collapse of the roof continues upward to the surface causing often swampy depressions in fields and woods. They look much like sinks in a limestone karst,

128. Road trench of old roadbed deepened by long years of erosion through the loess. This was long a dirt road. Gravel is a recent addition.

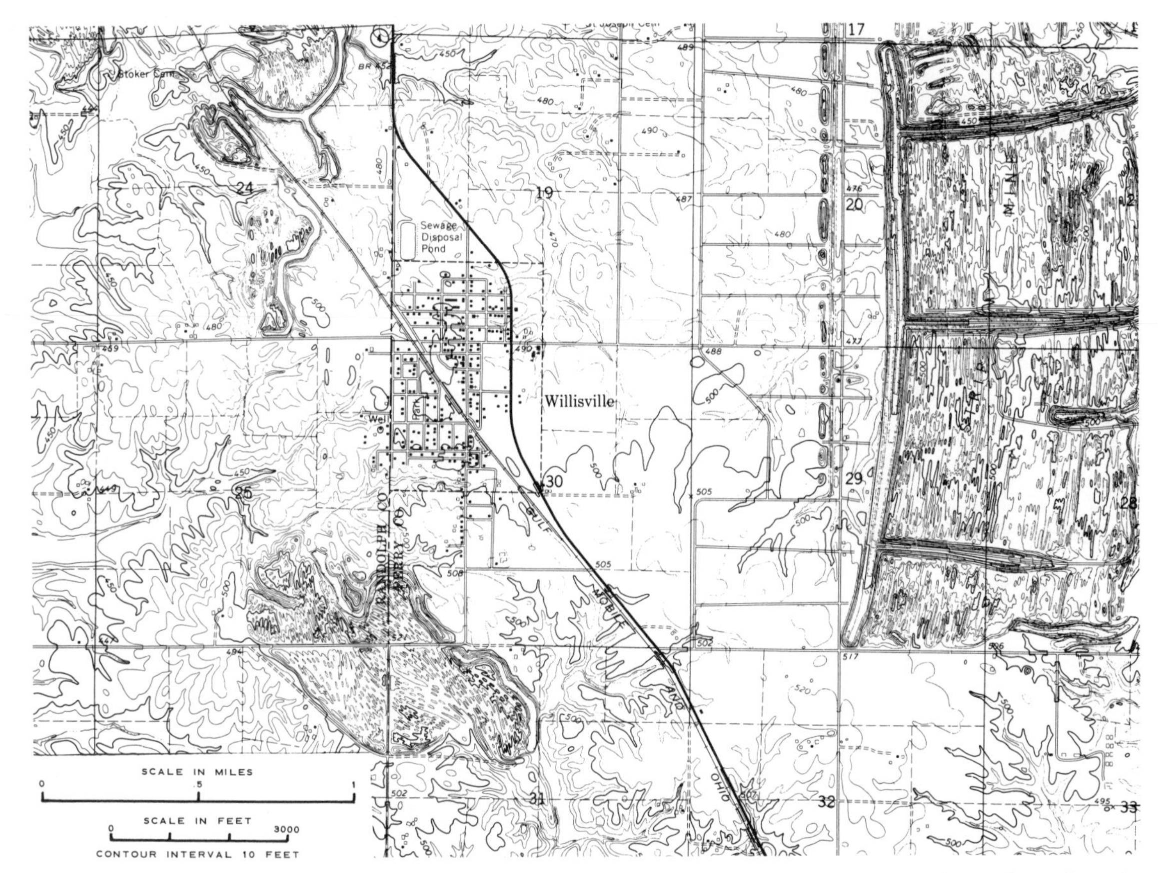

129. Map of the Willisville area, Perry County, shows a recent, rather orderly strip mine and more irregular strip mines of an earlier generation. The spoil areas of some modern mines are only gently rolling. At the west of the mine is a drainage ditch and its spoil piles.

130. Stripping operation in progress north of Du Quoin. Note the extensive leveled areas close behind the mining area. Dragline is at far end of working pit; stripping shovel is at bend in high wall.

131. An old strip mine shown from the air north of Du Quoin, Perry County.

132. Preparation plant and coal-storage piles west of Rend Lake, Jefferson County.

133. Eroding gob-piles and abandoned tipple north of Valier, Franklin County.

134. Aerial view of mine shown in Figure 133. *Although abandoned for many years, little vegetation grows on the gob piles and tailing flats.*

but they are usually arranged in a regular grid pattern. In towns, such mines have caused considerable damage to streets, houses, and water and sewerage lines, especially in Franklin County towns. It is presumed that the upward collapse will not reach the surface from the deeper mines.

Quarries reveal sections of rock back into the hillside or below the plain. Modern limestone quarries cover many acres (Fib. *135*). They may be worked by a succession of benches or a single face of one hundred feet or more. The quarry may be deepened as long as the quality of rock is satisfactory. Other limiting factors might be excessive groundwater flow or too great a depth to extract the rock economically. Many acres of land are required for crushing, storing, loading, and disposal of waste rock.

Most limestone quarries intersect cavities, especially near ground level. The cavities may be filled with residual clay and chert or be partially open and carry a stream of water whose volume varies with the season. Old quarries are interesting landscape features as vegetation clothes them, weathering releases fossils, and low places fill with water. Many small quarries from P.W.A. days (forty years ago) are interest points and no longer seem to be scars in the landscape.

Several limestone mines have been driven into the cliffsides along the Mississippi Valley, making use of high quality stone and excellent roof conditions (Fig. *136*). At Valmeyer, Monroe

135. A large quarry in St. Clair County. Successive benches are quarried out. The floor is below stream level, so water will fill the lower part after abandonment.

136. Pillars and entry ways of a large limestone mine (now abandoned) in Randolph County. A limestone of high calcium carbonate purity was mined here.

137. Dredge and barge of Southern Illinois Sand Company removing sand from the Mississippi River channel near Chester, Illinois.

138. Drainage ditch in flat terrace land, Gallatin County. Sides of ditch repeatedly slide in, causing gradual widening.

County, the mined-out portions are used to grow mushrooms and for high quality storage where temperature and humidity can be readily controlled.

Clay pits for brick and tile were formerly operated in many places. The abandoned pits remain, though overgrown and partly filled with water. They have slumped in and are not easy to see. A few pits continue to operate.

Sand and gravel pits are located along the larger rivers, especially on the terraces of the Kaskaskia and Wabash rivers. In addition, several companies dredge sand and gravel directly from the channel of the major rivers (Fig. *137*). The advantage of the latter operation is that the only land required is for screening and storage and loading; there is no abandoned pit. The river replaces the sand and gravel at the next high water.

Ditches groove the flatlands (Fig. *138*) and levees confine the stream course (Fig. *139*). On the uplands tiles have been laid to improve the land for cultivation. Old maps show wet and marshy lands which are now fairly well drained. In the lowlands stream channels have been straightened and deepened to carry off surface water faster. Levees have been built to reduce floods. The intended aims have generally been attained, but not completely. Wet years and heavy storms bring soggy ground and inundation, while straightened streams tend to promote channel deepening and cause gullying of the tributaries.

Channelization of the Kaskaskia and Saline

139. Skillet Fork southeast of Wayne City, Wayne County. Banks are "stabilized" only on one side in the foreground. Bends are shortened and straightened in the middle ground.

140. Relatively undisturbed channel of Kaskaskia River in foreground. Note small lakes and overflow channels. Fayetteville is in middle-ground, and the channelized river beyond.

rivers has greatly altered the landscape (Figs. *140* and *141*). The natural meandering is gone and the natural alternation of deep pools and shallow riffles is disrupted. The channel lacks deep refuges for fish in dry weather and turbulence to oxygenate the water. The channel has been widened to permit barge traffic, and the depth of the water is controlled by the locks.

Oil fields modify the land very little, but spills may affect the soil. The regular spacing of the wells, utility lanes, and access roads are inconspicuous identifying features (Fig. *142*).

Things to do. A search for cultural features of earlier days stimulates the imagination and gives insight into the past. Old maps and atlases can be perused at libraries and at historical society and county courthouse archives. Then a field trip in pursuit of the remains and topographic evidence bring challenge and satisfaction.

The magazine, *Outdoor Illinois*, has reproduced many such maps and contains many articles in this vein.

It is sometimes surprising that man's activities leave such a small mark or are soon so modified by natural processes that they blend into the landscape. Hopefully the greater impact of modern machines will also be ameliorated in time.

141. Dam and lock on the channelized Kaskaskia River near junction with Mississippi River.

SUGGESTED READING

Illinois State Geological Survey. Maps and reports of mines, quarries, sand and gravel operations. See especially Illinois Mineral Notes. Urbana.

Strahler, A. N., and Strahler, A. H. 1973. *Environmental Geoscience*. Santa Barbara, Calif.: John Wiley and Sons, Inc. 511 pp.

142. Agricultural land laced by pumps, pipelines and access roads, electric lines and storage tanks. Oil field in Hamilton County lies below this fertile soil.

Rocks

143. Alluvium along the bank of the Mississippi River in Alexander County deposited during high water. Erosion has exposed complex bedding features in the layers of silt and very fine-grained sand.

Sediments and Rocks

The rock beneath the Southern Illinois region is sedimentary. Characteristically it is layered into beds ranging from a fraction of an inch up to massive beds many feet thick. Sedimentary rocks are formed at the earth's surface under surface conditions of temperature and pressure, by means of the various chemical and physical processes which we are able to observe more readily than those which produced igneous and metamorphic rocks far beneath the surface.

By definition sedimentary rocks are derived from other rocks. Processes of weathering cause chemical alterations and solution, and mechanical disintegration into fragments. Gravity draws fragments downslope; running water, wind, and glaciers transport them to new places of deposition. As long as deposits rest above sea level they are subject to further erosion, but eventually the sediment comes to a resting place. So, too, dissolved materials may ultimately be taken from the solution by organisms, or chemical precipitation, or the evaporation of a water body. The deposits which result are sediments which can be dug with a shovel—gravel, snad, clay, lime mud —or perhaps be directly formed into rock as in a coral reef, or a crystalline deposit of salt or gypsum. Sedimentary rocks generally go through post-depositional processes of compaction, cementation, and chemical (crystalline) alterations.

Our sedimentary rocks are, in order of abundance, limestone, shale, sandstone, dolomite, chert, and coal. They were laid down layer by layer. The oldest ones, on the bottom of the pile, are all of marine origin. Rocks of the Mississippian and Pennsylvanian periods include sediments deposited in deltas, lagoons, swamps, tidal flats, and river channels and flats, as well as a variety of marine environments.

Deciphering the origin of the sedimentary rocks and the succession of geologic events which produced them is an exciting challenge to geologists; it is a game amateurs can play, too. Fossils are some of the best clues both to environment and to age. Other features such as ripple marks, cross-bedding, animal trails and burrows are equally valuable clues to environment.

Sediments rather generally cover the bedrock of Southern Illinois. River (alluvial) sediments fill or veneer the valley bottoms (Fig. *143*). Glacial deposits mantle the rock across almost all of the Central Interior Plains Division, and windblown silt (loess) lies above the glacial material and covers the uplands of the Ozark and Shawnee Hills divisions.

Sand, clay, and gravel sediments of the Coastal Plains Division across the south tier of counties have not been consolidated. They are distinctly different from the older rocks and from the glacial and alluvial sediments.

SUGGESTED READING

Illinois State Geological Survey. Urbana: Geological Science Field Trip Guide Leaflets.

Kuenen, P. H. 1960. Sand. *Scientific American* 203:94–110. (Offprint no. 803. San Francisco: W. H. Freeman.)

Pettijohn, F. J. 1949. *Sedimentary Rocks*. 3d ed. (1975). New York: Harper & Row. 736 pp.

———; Potter, P. E.; and Seiver, R. 1972. *Sand and Sandstone*. New York: Springer-Verlag New York.

Piskin, Kemal, and Bergstrom, R. E. 1967. *Glacial Drift in Illinois*. Urbana: Illinois State Geological Survey, Circular no. 416. 33 pp.

Willman, H. B. et al. 1968. *Bibliography and Index of Illinois Geology Through 1965*. Urbana: Illinois State Geological Survey, Bulletin no. 92. 373 pp.

———. 1975. *Handbook of Illinois Stratigraphy*. Urbana: Illinois State Geological Survey, Bulletin no. 95. 261 pp.

Sediments

Sediments are the products of erosion. They consist of the fragments of other rocks and also the precipitated materials which had been dissolved during the weathering process. In Southern Illinois all our sediments are fragmental. Precipitated materials do form stalactites in caves, the nodules and concretions found in loess and most other sediments, and the iron oxide partially cementing the Mounds gravels. When cementing materials bind the fragments fully, sediment becomes sedimentary rock.

Our sediments are the more recent deposits: the alluvium (river-laid sediment) and lake deposits, glacial and windblown materials, gravels found on the uplands, and the coastal plain sands, gravels, and clays of the extreme south. These lie above the Paleozoic sedimentary rocks which we speak of as "bedrock." They are classified on the basis of texture but also by geological origin. Thus, glacial till is a heterogenous mixture of sizes without bedding.

Gravels

Gravels refer to deposits in which the individual fragments are coarser than 4 mm. Gravels are very common in our area and occur in a variety of topographic situations. The stream gravels of the Ozark and the Shawnee Hills divisions are easiest to find since they are present in the channels of all the streams. Composition of the pebbles is mainly chert in the Ozarks, chert and sandstone in the Lesser Shawnee Hills, and sandstone in the Greater Shawnee Hills sections. These are modern gravels formed through weathering and mass-wasting of the bedrock. They are in the process of transportation downstream (Fig. *144*). The largest and most angular boulders are found in the source areas; breaking under impact and rounding by abrasion takes place rather rapidly.

Gravels are found associated with sands in all the bottomland provinces. Several companies dredge them from the rivers, screen them for size, and sell them mainly for construction purposes (Fig. *145*). Terraces along the Kaskaskia and Wabash rivers also yield gravels. These gravels consist of pebbles and cobbles originally transported by the Pleistocene glaciers. They contain pebbles of many kinds of rocks, especially cherts and crystalline rocks from the Canadian Shield. Most pebbles are well abraded and rounded.

Gravels and sands more directly associated with glacial deposits are found in the upland hills of the Effingham Plains section of the Southern Till Plains Division. Gravels are also found as masses and lenses within the main body of till in the Vandalia region and from place to place in coal strip mines. They are similar in nature to the glacial gravels of the bottomlands and terraces except that they are not so well stratified.

The high level upland chert (Mounds) gravels are the most intriguing geologically (Fig. *146*). They are most abundant on the ridgetops of the coastal plain region, but cap the Shawneetown Hills in Gallatin County and the hilltops close to the Cache-Bay creek lowlands and Thebes area in Alexander County. Scattered finds of similar brown chert gravels are known from Pine Hills, Monroe County, and large deposits are in Calhoun County. The gravels are mostly brown chert. When broken, the center of the pebbles are tan or white. The pebbles commonly have a high patina, especially in the concavities, but the patina is abraded from the rounded edges (Fig. *147*). Fossils in the pebbles are Paleozoic inverte brates, identifying the source as the Paleozoic limestones of the Central Lowland and Appalachian provinces. Quartz pebbles and large quartzite cobbles are also present. The Mounds gravels are associated with reddish sands, often in a matrix of clayey cement. They

144. Chert gravel bar in Hutchins Creek channel, Union County. Typical of the Ozark streams.

145. Mississippi River sand and gravel dredge near Chester, Randolph County. The gravels contain well-rounded igneous and metamorphic rock, and are largely reworked from glacial materials.

146. Brown chert gravels from an upland gravel pit west of Mounds, Pulaski County. Sand and clay fill the pores, and bedding is irregular.

147. Close-up of Mounds gravels showing the subangular shape of pebbles. Glint from surface of some suggests the distinctive polish on many of the pebbles.

are coarsely cross-bedded and have dense bands of dark brown iron oxide cement similar to that seen in the massive sandstones of the Shawnee Hills.

Sands

Sands are finer-grained fragmental sediments ranging from 2 mm. to $^{1}/_{16}$ mm. Their texture is coarse enough so that one can see the individual particles and feel them with the fingers.

Sands are commonly associated with gravels. They form sandbars in stream channels, especially in the large rivers. Sands make up levee and overbank deposits and the point bar deposits on the inside of meander bends. Sand is less visible in the smaller streams of the upland provinces where it is masked by the more abundant finer-grained silt. Most of the sediment removed from the Mississippi River by dredging is sand, though the finer grades are screened out and returned to the river.

Dune sands are common on the terraces along the Wabash (see Terraces). These sands are fine grained and better sorted than the river deposited sands from which they are derived. Figure *148* shows a stratified terrace sand whose exposed surface is being reworked and smoothed by the wind.

The coastal plain sands are very white, often micaceous and quite different in character from other sands of our area.

148. Sand pit at Roots, Randolph County where the Kaskaskia River enters the Mississippi Valley. Stratification is visible, but the surface has been reworked by erosion, especially by the wind.

Things to do. Each sand has its own set of characteristics. Note its relationship to the topography and to other sediments. Has it been recently transported? Is it a deposit left behind to give clues to conditions which no longer exist? Collect and label samples and then compare them, using a magnifying glass or microscope. Try to identify some of the minerals present and note the shapes and size ranges of the grains.

Silt

Silt is the texture next finer than sand. Coarse-grained silt looks grainy when dry, but it is so fine that individual particles are hard to distinguish. When damp, it is sticky and feels smooth to the fingers. Between your teeth it is gritty! When wet, it makes mud (Fig. *149*).

The surface materials from which our soils are formed are mainly silt. During Pleistocene glaciation the meltwaters brought tremendous quantities of sediment down the main river valleys. During summer months floods spread sediment on the floodplain. During winter months the source was frozen and the bare floodplain deposits were eroded by the wind. Vast clouds of dust carried the silt onto the upland where it was deposited on the bluffs and spread across the upland plains and hills. This windblown deposit is loess. From the time of deposit on the upland the loess was subject to erosion and subsequent deposition in stream valleys, lakes, and ponds, and even in caves. These later water-laid silt deposits are stratified, but some of the lacustrine layers are so evenly textured that bedding is not appar-

149. Silt deposited at the banks of the Mississippi River in quiet water as water levels recede.

150. Loess in road-cut near Belleville, St. Clair County. This very well-sorted silt was carried out of the Mississippi floodplain by winds. Near the valley the lower part is calcareous and contains abundant snail fossils.

ent. Silt usually forms the surface layers on the floodplains of small as well as large streams.

Loess

Loess is windblown silt which forms the surface material over most of Southern Illinois. It is especially prominent along the bluffs of the Mississippi River. Loess has the peculiar property of standing vertically for a long time (Fig. *150*). Road cuts in the East St. Louis region are especially striking.

Loess thickness is greatest on the bluffs and on the east side of valleys with wide floodplains. Above the American bottoms in St. Clair and Madison counties wells have drilled through more than fifty feet of loess. Thickness decreases eastward.

The texture of the loess is also coarsest near the source and may even be somewhat sandy. On the bluff, too, there is much calcium carbonate so that a drop of dilute hydrochloric acid will cause effervescence. Snails, very much like the modern woodland snails of the region, are embedded in the sediment. Nodules of calcium carbonate are common. They take interesting and peculiar shapes. Long ago the nodules were named "loess kindchen" —children of the loess—by Germans of the Rhine Valley where similar deposits occupy the same topographic position. A few miles away from the valleys the loess is no longer calcareous and neither fossils nor nodules are present.

Things to do. Look carefully at an exposure of loess. You will probably discover that the lower part is redder than the buff color so characteristic of the surface silts. These represent two separate deposits. A weak soil is present between them. At the bluff this buried soil zone includes a few feet from which the calcium carbonate had been leached away before deposition of the later loess. Fossils are found only in the calcareous portions. Collect the fossils from the upper loess and contrast them with the assemblage from the lower loess.

Clay

Deposits of true clay are not common in Southern Illinois. Clay is so fine grained that it feels smooth as chocolate between the teeth. Silt impurities would give some grittiness. There are several varieties of clays which have very different properties. In general, clay swells when wet but dries and contracts becoming hard as a brick. Clays tend to be chemically active, taking up, holding, and releasing plant nutrient ions.

Clay deposits of the Coastal Plain are the thickest and most interesting. A clay pit near Olmsted, Alexander County, has been worked for many decades, though by different companies and for different purposes. Several small deposits of kaolin clay-filled sinkholes are found west of Cobden, Union County. In 1975 the pits were full of water and the clay was hard to find. Prospect drilling indicates that there are other small deposits in the area.

Clays are deposited in backwater areas of floodplains along the major rivers. They may partially fill abandoned meanders and other low places. Soils maps identify their location, and soils descriptions tell about their properties.

One of the most interesting occurrences of clay is in caves. "Cave clay" is orange to dark red and buttery smooth. It has a distinctive smell. These clays predate the silt and gravel fills. They are found in pockets and on shelves and sometimes in banks of some size in areas which were protected from erosion by underground streams. The cave clay came from the limestone. It was an insoluble impurity left behind by the dissolving of the limestone. The deposit must represent the residue of many feet of rock. The clay particles were not eroded away by running water, rather they were carried downward through crevices in the rock and accumulated in the slowly enlarging caverns below ground.

Glacial Till

Glacial till is the heterogeneous deposit left in place by the melting away of glaciers (Fig. *151*). In Southern Illinois it us finer-textured than the tills of central and northern Illinois. Few boulders are enclosed within it. Our tills are mainly of silt intimately mixed with clay and sand. In many places the boundary is hard to draw within a foot or so between the silty soil zone of the till and the base of the overlying loess. Probably some mixing by plants and animals took place as the windblown silts gradually accumulated on top of the till soil.

The key to identification of till is the pebbles that are enclosed. Most pebbles are likely to be sandstone which, like the silt and clay, came from the Pennsylvanian bedrock as the ice scraped across it. Also prominent are white chert pebbles, not found in Pennsylvanian rocks. A little search reveals crystalline rocks—both igneous and metamorphic. Pebbles, even cobbles, of the granite family, basalts, gneisses, and greenish amphibole and chlorite schists are all foreign to our region; they must have come from the Canadian Shield north of the Great Lakes.

Glacial till lies beneath the loess and above the bedrock throughout the Southern Till Plains Division. It is also found in the northern section of the Ozarks and along the northern slope of the Shawnee Hills (see Fig. *1*).

151. Close-up of till on the side of an excavation. Raindrop erosion has left the coarser fragments on the surface in upper right. The ditch exposes several larger pebbles and a polygonal joint pattern in lower right. The joints were probably formed by drying and wetting.

Sandstone

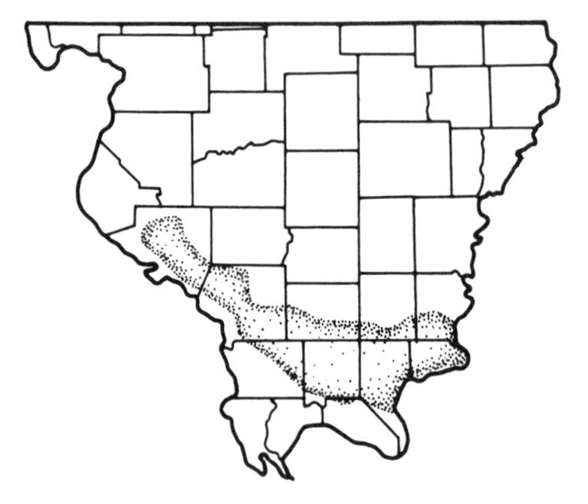

A sandstone rock looks and feels grainy. The texture of the grains range from 2 mm. down to 1/16 mm. This is rather fine, but individual grains can be distinguished with the naked eye. The common color is yellowish brown on exposed surfaces, but lighter when first broken. Chips of rock from wells drilled below the water table are gray; thus the yellow color is due to the oxidation (rusting) of iron. Red sandstones are common, but very pure, white quartz sandstones are rather rare. The composition of the sand grains is mainly quartz. This is true of all the older Paleozoic sandstones up through the Caseyville (see Appendix C for formation terminology and relationships). In the upper Pennsylvanian sandstones, especially upward from the Murphysboro Coal, increasing numbers of grains are feldspar. Many bedding surfaces are shiny with mica flakes.

A magnifying glass or low-power microscope reveals the nature of the sand grains themselves. The grains of the Devonian Dutch Creek sandstone up to the Ste. Genevieve Formation are white, very well rounded, frosted and pitted. Above the Ste. Genevieve the grains are mostly angular (Fig. *152*). Commonly a secondary silica growth has re-formed crystal faces. In extreme cases the silica on adjacent grains has grown together forming a dense and resistant rock. When such sandstone is broken the fracture breaks right through the sand grains as it does in quartzite.

Sandstones are layered or bedded. Massive layers (see Fig. *153*), eight feet or more in thickness, form prominent cliffs. Vertical fractures or joints were produced by stresses exerted by the gradual warping of the earth's crust. The thicker the layers the more widely spaced the joints. On hillslopes gravity may cause great joint-bounded blocks of sandstones to slide away, forming a landscape like Jackson Hollow, Pope County, and Flannigan Township, Hamilton County, and hundreds of other scenic beauty spots. Thin-bedded sandstones (see Fig. *154*) have closely-spaced joints forming flagstones and steplike sheets of rocks exposed in stream beds and rock cuts at thousands of places across the Shawnee Hills.

Thin-bedded sandstones often have partings of clay or shale. Shale and sandstone are closely associated so that the presence of one suggests that the other will be nearby.

Exposed surfaces may have curious and intriguing features. Prominent on massive cliffs and shelter bluffs are pockets, scalloping, and boxwork. The protruding ridges are dark brown concentrations of iron oxide cement. Boxwork is caused by concentrations along intersecting joints. Many of the pocked surfaces have no apparent structures to account for the irregular distribution of cement. One explanation is that organic matter, once at the center, caused precipitation of iron oxide around it. Later when erosion exposed the rock near the surface, groundwater caused outward diffusion of the iron oxide from the center, producing a shell of concentrated cement and leaving a center uncemented. On exposure the grains readily wash away. In freshly exposed places grains can be easily rubbed apart.

Sandstones are the most resistant rocks in our region, so they cap the ridges and high

152. Coarse-grained sandstone from the Caseyville Formation. The grains are mainly quartz, angular to subround. A few grains show overgrowths of additional silica producing crystalline faces.

areas. They are the cliff formers and the cap-rock of waterfalls. Because sandstones are resistant to weathering and erosion they are the most visible rock.

Many depositional features are preserved from the time of their origin. They are the clue to fascinating interpretations of past environments. Look for ripple marks, cross-bedding, burrows and tracks, current scour, plant or invertebrate fossils, mud cracks, and raindrop impressions. (See sec. Primary Sedimentary Features.)

153. Massive sandstone forming cliff face. The even vertical surfaces are joint surfaces that cause large blocks to fall when erosion leaves them unsupported.

Shale

Shales are fine-textured sedimentary rocks. The term "shale" implies a tendency to break with a platy fracture. They are weak rocks subject to rapid disintegration when exposed to weathering. Some expand when wet and the soft clay minerals become slippery. Clay banks crumble and slide and running water easily transports the fragments.

Natural exposures of shale are not common even though the total thickness of shale layers is greater than sandstone in our region. Natural exposures are found beneath waterfalls and where stream channels swing against the outcrop on the valley sides. Road cuts, quarries, strip mines, and other man-made excavations give the best exposures.

Shale is associated with sandstone and limestone. When it is associated with limestone look for marine invertebrate fossils which may be entire carbonate shells or merely impressions in the rock. When shale is associated with sandstone the original environment may have been nonmarine and plant fossils occur.

Shales have great variety (Figs. *155* and *156*). The coarsest textures are siltstones, and grade to sandstone, while the finest claystones are slick and soapy to the touch and individual particles are crystals of clay minerals visible only with an electron microscope. Most shales include both silt and clay particles. Some idea of the range can be determined by shaking up a handful of crushed shale in a quart jar filled with water. Coarse silt will settle rather rapidly, but colloidal clays keep the water cloudy.

Shale is most commonly gray in color. Look for fossils and test the shale with a drop of dilute hydrochloric acid to see if it is calcareous (limey). If calcareous it is probably marine and may enclose fossil shells. The fossils tend to be concentrated in certain layers. Nonmarine gray shales may contain plant fossils. Carbon films of ferns may be beautifully preserved. Extraction may be difficult, though, because the shale disintegrates, breaking the fossils into hundreds of pieces. Concretions, which are ovoid cementations of the shale around a

154. Siltstone and shale at spillway of Kinkaid Lake, Jackson County. Note the many flat, platy fragments loosened in the collapse block; these are subject to erosion in the next storm.

155. Clayey shale bed beneath a massive quartz siltstone at Coles Mill Section, Chester, Randolph County.

156. Shale containing abundant fossils interbedded with lenses of limestone that are also fossiliferous, Illinois 146, Union County.

plant fossil, provide the most satisfactory preservation.

Black shale is associated with coal, has much carbon in it, and probably some brassy-looking pyrite (FeS_2). Look for invertebrate pyritized fossils, especially Pecten clams (the Shell Oil Co. symbol) and small Brachipoda. The black shales (slate to coal miners) commonly lie directly above coal beds, and represent lagoonal conditions, transitional between the coal swamp and the sea.

Red or pink shales are generally blocky rather than platy. They are not so common in the Pennsylvanian rocks of Illinois as they are in the western United States. In Southern Illinois the Ste. Genevieve and upper Chester formations contain dark reddish shales associated with reddish fragmental limestone. The most accessible exposures are road cuts on U.S. 51 one mile north of the intersection with Illinois 146, and on I-57 just north of Big Creek bridge (Fig. *165*).

Conglomerates

Conglomerates are made up of gravel-sized pebbles cemented together. Finer material, usually sand, fills the spaces between the pebbles. Conglomerates are not abundant in Southern Illinois but where found they are spectacular rocks.

157. A cobble of quartz-pebble conglomerate from the Caseyville Formation. The largest pebbles are about ¾ inch long, smooth and rounded. Such concentrations of pebbles are not seen in many places; one such place is Draper's Bluff, Union County.

The pebbles which make up a conglomerate have been eroded from some other rock, so they have been transported, bumped, and abrades. Some pebbles are smooth and round, others very angular. The composition of the pebbles might be that of any rock weathered, broken, transported, and reincorporated. Quartz pebbles are most common in the massive sandstone units of te Shawnee Hills. True quartz-pebble conglomerates are rare, but pebbly (conglomeratic) sandstones are common in the Battery Rock and Pounds sandstones (see Fig. *157*). The source of the quartz pebbles is unknown. They do not derive from the underlying Chester rocks nor from any older rocks in the Central Interior Plains. We can only speculate that they came from the Appalachian or Canadian mountains to the north and east.

Shale pebbles are most frequently found in the upper Ste. Genevieve and the lower Pennsylvanian units. The pebbles are generally disc shaped, up to two inches or so in diameter. The shale pebbles are quite fragile, easily split by wetting and drying and freezing and thawing. Holes in the exposed rock faces may be all that is left to record their presence; break off a fresh slab and whole pebbles will be found. Shale is so fragile that it will not stay together for long during transport, especially since these must have been dried clay fragments picked up from a mudbank during a storm and deposited with the sand grains. The source beds were therefore nearby.

Limestone pebble conglomerates are uncommon in the Southern Illinois counties. The most striking is a thin unit visible at low water in the channel of the Mississippi River about a mile north of Thebes, Alexander County. The limestone pebbles are slabby in shape, up to five or six inches across with angular edges. These, too, had a local origin. A deposit of fine-grained mud apparently dried out, and was then ripped away by a storm and enclosed in more lime mud. This is a unit of the Girardeau Formation of the Silurian period. Another famous locality of similar but more rounded limestone conglomerate is found in Calhoun County in Irish Hollow near Hamburg.

Cherty conglomerates are found near the base of the massive Pennsylvanian sandstone units in many places across the Shawnee Hills. Their distribution is very local. The pebbles are angular, white or gray in color, and up to two or three inches across. These pebbles generally lie on an unconformable surface, perhaps in a scour channel later filled by the conglomerate and sandstone. The chert pebbles are derived mostly from erosion of the cherty limestones of the Chester Series. Near Alto Pass and on Fountain Bluff chert pebbles with Devonian age fossils have been found. They lie just north of the Rattlesnake Ferry Fault which has Devonian cherty rocks on its south side.

Limestone

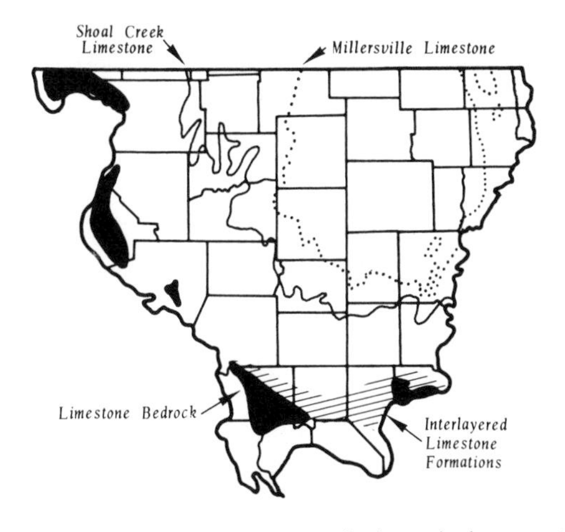

Limestone has many varieties. It is most often light gray, compact, and hard to break. It is easy to scratch with a knife but does not have a soapy feel. It is composed of calcium carbonate. A drop of 10 percent hydrochloric acid will cause effervescence (bubbling) (Fig. *158*).

Limestone is best exposed in limestone cliffs, in quarries, and road cuts (Fig. *159*). The index map shows the distribution of limestone bedrock outcrop. Weathered limestone looks different from a freshly broken surface. In the field it is usually white or very light gray. Weathering has etched out the more soluble material often leaving fossil fragments and slightly more resistant layers standing in relief. Flint nodules, lenses, and layers are also prominent. A freshly broken surface is generally darker than the weathered surface. A crystalline variety sparkles from the cleavage surfaces of the calcite, while a fine-grained rock appears smooth and dense and dull.

Fossils are common constituents of limestone (Fig. *189*), though they are most easily collected from impure limestones from which they may weather out in their entirety. Some limestones consist of fossil fragments which were spread over the sea floor by waves and currents. Fine-grained lime mud filled the space between fragments. After burial recrystallization may obscure the original nature. A

158. Thin-bedded but dense limestone at one of the entrances to Krueger Cave in Monroe County. Note how the more resistant layers protrude; the more soluble or clayey layers form reentrants in the rock wall.

broken surface looks merely crystalline, but weathering selectively reveals the fossil fragments.

Some limestones are very dense and fine textured. When broken the fracture is conchoidal, or smooth and curved like a piece of broken glass. The very finest grain is called "lithographic" because it is used to make lithograph prints and the finest maps.

Impure limestones are common. Clay and silt may be in very thin lamination or their particles may be distributed evenly among the carbonate particles. All the Pennsylvanian limestone formations and most of the Chester limestones have such impurities. Entire fossil shells may often be collected from the weathered exposures of these rocks. The lower Paleozoic limestone formations tend to be purer carbonate. On the other hand some, especially the Silurian and Devonian limestones, have disseminated silica, which makes them more resistant to weathering. Interbedded layers and nodules of chert are common in these older limestones (Fig. *160*).

In humid regions limestone tends to be dissolved by surface runoff water and groundwater. The solution causes pitted surfaces and cavities, especially along joints and bedding planes (Fig. *161*). Road cuts, quarries, and cliffs reveal these in cross section. Springs may issue from them or the cavities may be nearly filled with clay which has worked its way down from the surface. Typically, limestone terrane is a lowland or an upland plateau entrenched by one or more major river valleys. High, stable limestone cliffs form the valley walls, as at the Pine Hills and along the Mississippi Valley in Randolph, Monroe, and St. Clair counties. The limestone upland may have sinkholes, disappearing streams, caves, and springs. Karst topography develops where the limestone is pure and thick and dense. Karst features are less prominent where the limestone has impurities or interbedded layers of shale or sandstone.

The soils developed on limestones are typically clayey with much chert or siliceous residue. The limestone itself does not become part of the soil because it is dissolved away. Only the insoluble materials remain behind.

In the Ozark Division many of the limestones have been altered to chert. The fossils, originally shells of calcium carbonate, are now silicified. Other primary features, characteristic of limestone, are preserved, but the rock is altered.

159. Dense, gray limestone in quarry at Dupo, St. Clair County. Solution cavity, mostly filled with silt, yields a small stream of water.

Chert (Flint)

The terms chert and flint are synonymous. Chert is used more generally by American geologists. Chert is a microcrystalline siliceous rock associated with limestone. It ranges from white to black, and brown shades are common. Red chert (jasper) is not native to Southern Illinois but has been found in glacial deposits of till and gravel. Jasper pebbles are not uncommon in Mississippi River gravels.

Chert is nearly as hard as quartz and will scratch a knife blade. Although hard to scratch it is brittle and shatters under a sharp blow. Heat also causes shattering because some water is contained within minute openings. Indians chipped the edges of their flint implements by this means. By the same token it is dangerous to put chert rocks around a camp fire because explosions project sharp fragments with a good deal of force.

A broken surface has a smooth rounded surface (conchoidal), like broken glass (Fig. *160*). The texture may be as fine and shiny as porcelain or more grainy. The luster is generally described as porcelaneous and certainly is more reflective than the limestone associated with it. Perhaps that feature is the most distinctive for recognition. Then check its hardness. It will not effervesce in hydrochloric acid, but sometimes calcareous fossils or crystals enclosed in the chert will effervesce so be careful when using the acid.

Fossils in limestones have sometimes been selectively replaced by silica. Such fossils remain on the surface of a weathered exposure. In some siliceous limestones, especially in the Devonian rocks of our area, delicate spiny

160. Black flint nodules in limestone riprap at Rend Lake spillway, Franklin County.

shells can be separated from the rock by putting a block in an acid bath and dissolving away the enclosing limestone. Natural weathering processes do not preserve the delicate parts.

Chert occurs in layers up to several inches thick. These may be quite extensive, traceable for many yards in a quarry face. Smaller flattened or even dumbbell-shaped nodules are also common. Generally they lie at rather specific horizons. Some formations such as the Devonian Grassy Knob and Clear Creek of southwestern Illinois are extremely cherty, others have abundant chert layers such as the St. Louis Formation in Union County. The Salem and upper part of the Ste. Genevieve Formations have almost no chert.

Chert is undesirable in concrete aggregates because of its shattering properties, but is acceptable in the crushed rocks used for road gravels and parking areas.

A special variety of chert, called tripoli, has a very grainy appearance. It is white, though sometimes stained buff by iron oxide. Tripolitic chert often forms a rind on the outside of chert nodules, but in Alexander and Union counties considerable thicknesses of tripoli make mining possible. The cottony looking rock scrapes apart easily and one might think it soft until attempting to polish a knife blade on it. Then the hardness can be felt and heard. Commercially this chert is called silica. It is ground and milled to powder of various textures and sold for its abrasive or resistant properties. It has been used as a substitute for lime in lining football fields because it is inert and does not burn the skin when wetted.

Chert pebbles dominate the gravels of streams where bedrock is cherty limestone. This is particularly true of the Ozark Division of southwestern Illinois and throughout the Missouri Ozarks. High-level brown chert gravels between Cache bottoms and the Ohio River represent a time when a broad alluvial plain covered the head of the Mississippi Embayment some three hundred feet above the present level of the river plain.

161. Solution by ground and surface water has widened crevices, rounded and pitted the surface of a limestone exposure washed by the water of Hutchins Creek, Union County.

Coal

Coal in Southern Illinois is bituminous. It is black and prominently banded with alternating shiny vitrain layers, dull durain, and grainy fusain layers. The vitrain is hard but the fusain is soft, dirties the fingers, and is mainly responsible for the dust and the dark smoke particles. Bituminous coal is coarsely blocky because of joints, or cleats, at right angles to each other (Fig. *162*).

All the commercial coal in Illinois is of Pennsylvanian age, but thin beds of limited distribution are found in the Chester Series. In Southern Illinois the Herrin No. 6 bed is most widely mined, but the Harrisburg No. 5 (which lies from eighteen to one hundred feet below the No. 6) is nearly as important. Their band of outcrop is shown in Figure *2*. The tremendous swamps in which vegetation grew and was preserved, extended from Peoria to western Kentucky and far to the northeast. How much farther the swamps may have extended toward the southwest we can hardly surmise because erosion truncates them along the outcrop margin. Actually the swamp may not have existed over the entire area simultaneously. Rather, it may have expanded in one direction while being gradually filled and covered by other sediments in its older parts.

Coal was formed in a coastal plain environment. Associated with it are thin shales, sandstones, and limestone. Some are marine, some nonmarine. Fossils contained in the rocks tell much about the environments in which they were deposited. Commonly this association of rocks occurs in repeating cycles, suggesting successive advance and retreat of the seas.

Southern Illinois coals have a reputation for a high sulphur content. The visible sulphur is in the form of iron sulphide (FeS_2) and occurs in two crystalline species, pyrite and marcasite. Both are brassy in color and form attractive crystals. These forms are in thin sheets along the cleats of the coal or in large nodular masses. Most of the pyritic sulphur can be physically separated and removed during the washing process. Chemically bound sulphur is

162. A blocky piece of bituminous coal. The shiny, black and hard vitrain layers show clearly. The white material along the vertical cleats are impurities that will become part of the ash.

much more difficult to remove, since it was taken up by the plants and formed organic compounds. Sulphur in this form must be removed from the stack during burning.

Coal is not easily found in the outcrop because it is usually covered by the weak shales which commonly overlay it. Strip mining has removed or covered many potential natural exposures, but, on the other hand, it gives an unparalleled opportunity for detailed observation of the coal and rocks. Horizontal variations can be mapped and studied along the mine high wall, but this is so soon removed that the work must be done with speed and care.

SUGGESTED READING

Hopkins, M. E., 1975. *Coal Mines in Illinois* [Map, 1:500,000]. Urbana: Illinois State Geological Survey.

Smith, W. H., and Stall, J. B. 1975. *Coal and Water Resources for Coal Conversion in Illinois*. Urbana: Illinois State Water Survey and State Geological Survey. Cooperative Resources Report no. 4. 79 pp.

Igneous Rock

Igneous rocks in Southern Illinois! From glacial deposits, of course, but many dikes and sills have been found in the southeastern counties. The more common dikes intersect the horizontal sedimentary layers like a vertical wall, while sills are irregular sheets nearly

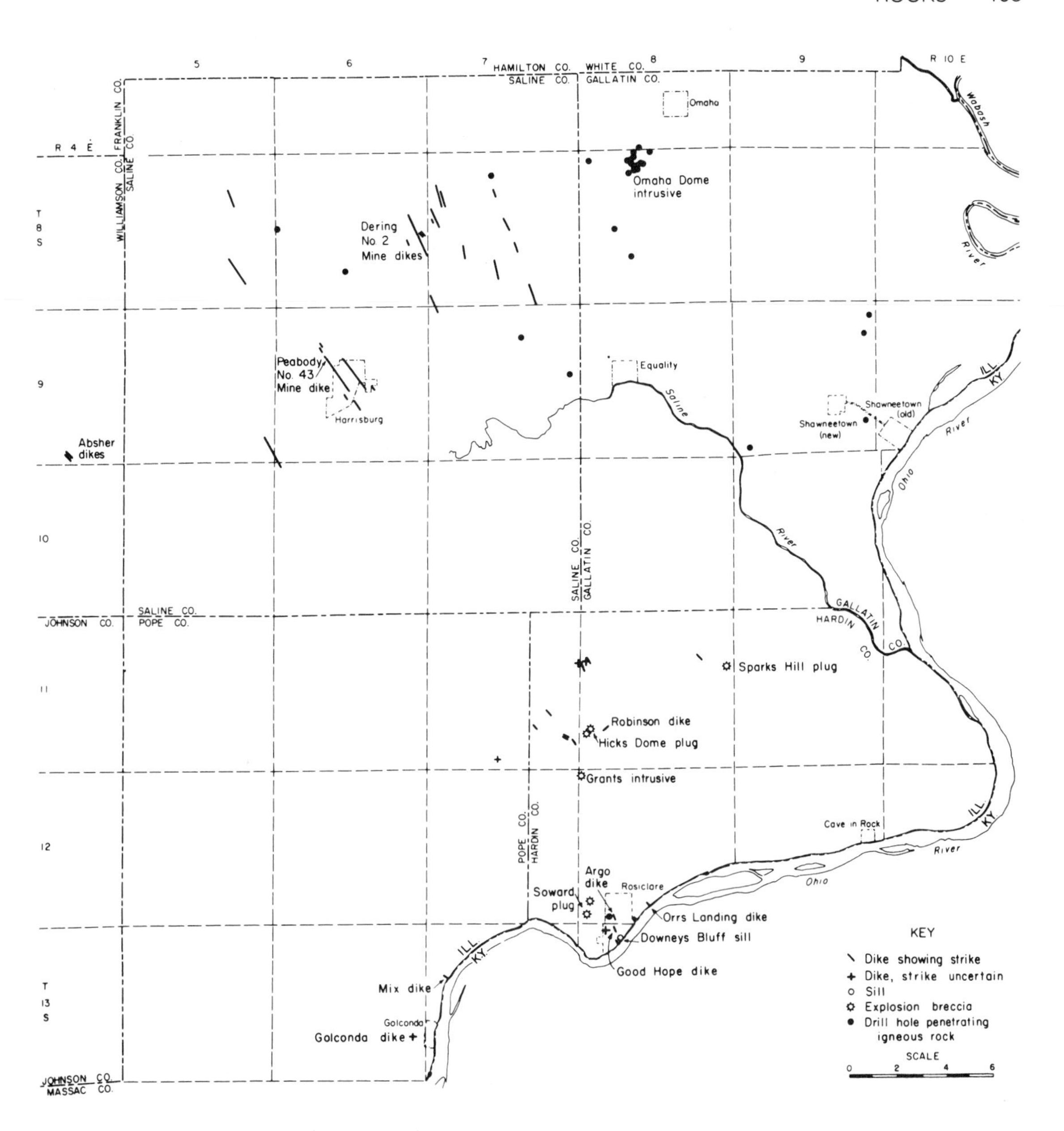

163. Distribution of igneous intrusive rocks in Southern Illinois. Some occurrences within city limits of Rosiclare not shown for sake of clarity (from Illinois State Geol. Circ. no. 330).

parallel to the bedding. The largest known exposure, uncovered in a coal strip-mining operation in Saline County, was three hundred feet thick. All the known natural exposures are in Pope and Hardin counties (Fig. *163*). Well drilling and underground mining have encountered similar igneous rocks in Gallatin, Saline, and eastern Williamson counties.

The dike rocks are all dark gray or greenish gray and similar in character. Large crystals, or phenocrysts, are enclosed in a fine-to-medium crystalline matrix. The primary minerals are dark mica, pyroxene, olivine, apatite, magnetite, and titanite. These have been much altered into carbonate, serpentine, and chlorite, so that some intrusives hardly look like igneous rocks. Another peculiar feature is the many enclosed fragments of sedimentary and igneous rocks and particles, including sandstone, limestone, shale, chert, slate, and granite. Some of the rocks are so fragmented that they are considered to be explosive breccias similar to peculiar structures of "jumbled" rocks found in several other places in the Midwest.

These igneous rocks are not easy to find because exposures are small and the rocks are weak. A few at known locations are accessible. When walking and exploring in the southeastern area keep your eyes open for discordant, dark rocks that intersect the normal sedimentary layers. A new discovery would be a valuable addition to geologic knowledge. Be sure to locate it accurately and, if possible, take a photograph and a sample specimen to show to a professional geologist.

Igneous intrusions can be recognized in the field principally by their relationship to the enclosing sedimentary rocks. Generally the igneous rock is more weathered, thus making a hollow, whereas the sedimentary rock adjacent to it was metamorphosed and is probably more prominent. The coal adjacent to an igneous intrusion has been heated so as to drive off the volatiles and form natural coke. The igneous rock is likely to have quite a different color, texture, and jointing pattern from the related sedimentary rock. The broken rock fragments enclosed in many of the igneous rocks are so unusual that attention is drawn to it.

SUGGESTED READING

Clegg, K. E., and Bradbury, T. C. 1956. *Igneous Intrusive Rocks in Illinois and their Economic Significance*. Urbana: Illinois State Geological Survey, Reports of Investigations no. 197. 19 pp.

Primary Sedimentary Features

Sedimentary rocks display many features similar to those which can be observed in modern sediments. The modern features can be related directly to the processes forming the sediments and to the ecologic environment surrounding them. It is therefore possible—although not without some disagreement—to reconstruct the sedimentary environments of our region when those rocks were formed. When present, fossils provide additional assistance.

Sandstones almost always have intersecting features, both in cross sections through the beds and on bedding surfaces. Upper and lower surfaces have distinctly different appearances.

Cross-Bedding

Cross-bedding is the most noticeable sedimentary feature since it occurs in the mas-

164. Fine-grained sand deposited in the spring floods of 1973 near the Ferry Landing, Grand Tower, Jackson County. The cross-bedding shows that the current moved from left to right. The beautiful etched form is due to drying and reworking by the wind. The finer layers are more cohesive.

sive sandstones exposed throughout the Shawnee Hills and in the rocky areas of the Mt. Vernon Hill Country (Figs. *164* and *165*). There are many forms of cross-bedding which are caused by deposition of sand from a bed load being moved by currents of water or by wind. Velocity of the current is decreased behind an obstruction or in a depression, causing deposit on the lee side. The resulting layers slope in the direction of the current flow. Small ripple marks, large sandbars, the fill of scour channels, and windblown sand dunes are all cross-bedded. The photographs show generally horizontal surfaces of deposition with sharply dipping layers between.

Scour Channels

Complex, parallel, elliptical features may be found on a large exposed surface of sandstone (Fig. *166*). The current was parallel to the axes of scour fills. The direction of slope of the cross beds within the scour and the steep side of the ripple marks on the surface point the direction. Careful examination of the cross beds reveals a dishing shape caused by sand rolled down the sides as well as the end of the scour (Fig. *167*). Once filled with sand the top beds have a horizontal orientation and the upper surface is commonly ripple-marked.

The scour channels may be only a few yards

165. Cross-bedding is clearly displayed in the resistant layer at the base of the road-cut on new U.S. 51 east of Anna, Union County.

166. Elliptical scour channel fills in massive sandstone at Kerr Bluff, Union County. High velocity currents scoured out the sand, then decreasing flow channeled sand to fill the scour with cross-beds. When filled the upper surface was ripple-marked. Successive episodes of scour and fill marked the deposition of this rock some 300 million years ago.

167. Cross-bedding in foreground and middle ground in fine-grained sandstone, Kinkaid Lake spillway, Jackson County. Dishing of the cross-beds can be seen at left.

long. They were caused by floodwaters moving at high velocities. Scoured sand was picked up and transported in suspension. As current subsided, the sand dropped to the bottom where it was swept along until it rolled down into the quieter water of the depression. There it stayed and was finally buried.

Ripple Marks

Rhythmic waves of sand are called ripple marks. As in the case of cross-bedding, they have many forms and variations which help in the identification of the ancient environment (Fig. *168* illustrates two forms on successive beds). Asymmetrical ripples indicate the current direction, with the steeper depositional slope representing the down-current side. Ripples may form long ridges which represent moderate velocities and an abundance of loose sand. Faster currents with more turbulence produce small cups with an arc-shaped ridge around them. These are referred to as interference ripples.

Ripple marks are familiar features in modern environments such as a sandy stream channel (Fig. *169*) or a beach or sand dune. They are rhythmic ridges of sand. It is interesting to watch sand being carried up the flanks of a ripple, roll down the steeper down-current side, and stop in the quieter water of the trough. After a time one realizes that the upstream side is being eroded and that the entire ripple moves down current. The form and shape shifts and changes. In sandstone we view suspended motion, when the current died

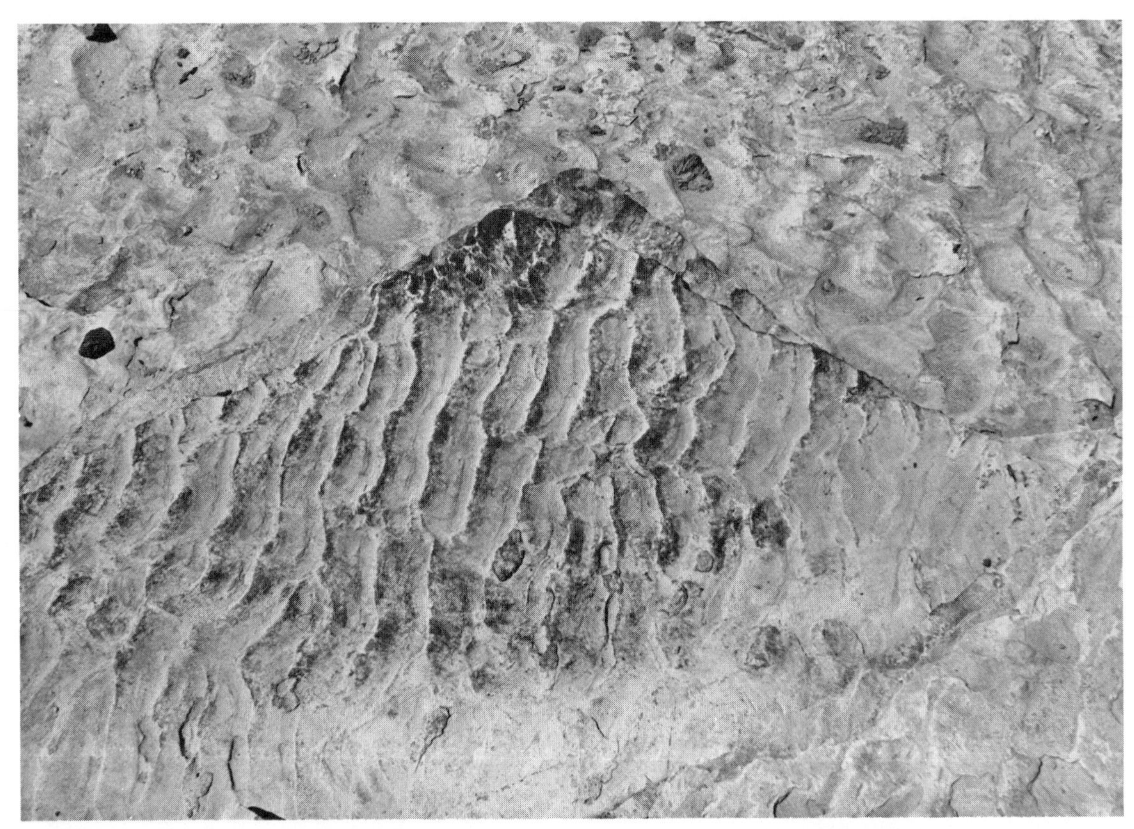

168. Ripple marks are common features on the surface of Pennsylvanian and Chester sandstones and in the sandy channels of modern large rivers and small streams. These are in rock, and were formed some 300 million years ago. Direction of movement was toward the left. The linear ripple marks were formed by currents of moderate velocity, while the arc-shaped interference ripples were formed at higher velocities.

away and the next surge of transported sand buried and preserved that which went before (Fig. *168*).

Laminations

Laminations are very thin parallel layers of slightly different textures of fine-grained materials. They commonsly represent deposition in very quiet waters. Laminations are common in dense limestones such as the Sexton Creek Formation near Thebes, Alexander County, and the St. Louis Formation in Monroe and St. Clair counties. Some coarse laminations may be seen in the top layer of Figure *169*.

Mud Cracks

Mud cracks are due to the contraction of fine-grained material as it dries. They are seen on river banks, the exposed floor of a drying pond, and in dried deposits of rock quarries and clay pits. Occasionally, they may be found in siltstones and shales, but these rocks are rarely well exposed.

When you find mud cracks, pick up one of the polygons and see how finely laminated it is with fine-grained fragments of different textures. The polygon separates along a coarse-textured layer. The cracks themselves may penetrate several inches downward. Many

169. Thin layers of silt and sand deposited at floodtime by the Mississippi River at Thebes, Alexander County. Subsequent erosion reveals the intricate patterns of deposition. Most prominent are the "climbing ripple marks," but fine cross-bedding can be seen in the lower layers.

preserved cracks became filled with other sediment before burial.

Mud Curls

When fine-grained deposits are thoroughly dried out, thin layers of clay or very fine-grained silt resting on coarser material may curl up and make a roll. Wind may then carry it away. Broken mud curls are sometimes incorporated in other deposits.

Raindrop Impressions

Raindrop impressions are commonly associated with mud cracks (Fig. *170*). They represent continental or tidal environments where there is alternating wetting and drying. The impressions must represent a sprinkle, at a time when the mud was still soft. Preservation requires that the surface of the mud must subsequently have become hard and dry and then be rapidly covered by the next sediment layer. Such impressions may be seen split apart by mud cracks.

Load Casts

The underside of sandstone layers is often rough and irregular, especially where the sandstone layers alternate with thin silt or clay layers. Load casts are knobs and ellipsoids,

170. Raindrop impressions in the mud along a riverbank. Splashes from large drops make a crater the size of the penny, but the whole surface is marked by small depressions.

with a somewhat streamlined form. They are apparently fillings of slight depressions in the underlying layers, followed by downward squeezing of more sand due to the increasing weight or load of accumulating sediment above.

Burrows

Thin-bedded sandstone units often contain the filled burrows of bottom-dwelling animals. The burrows displace the original sand. In some cases an open burrow is filled with somewhat different material. These usually have sharp, vertical boundaries and rudely layered filling. Horizontal burrows made just beneath the surface are more common. They are in many patterns. They, too, may have sharp boundaries and curving ridged and grooved material within them. They often intersect.

Burrow fillings are often found in Pleistocene and recent silts. Most appear to be crayfish burrows that were made in such wet places as waterways and floodplains.

There are many other internal and surficial features of sediments and sedimentary rocks. These are described in detail in geologic textbooks and journals.

Things to do. Go to a place, such as a stream bed or lakeshore, where sediment has recently been deposited and exposed by the lowering of the water level. How many of the previously described features can you find? There will be many more—grooves made by a floating log, for instance. Can you figure out their origin? Compare these features with those you find in sedimentary rocks. Since the rocks were formed on an inland sea floor and in deltas building into it, as well as in river channel and overbank deposition, there will be many differences.

SUGGESTED READING

Allen, John R. L. 1968. *Current Ripples*. Amsterdam: North Holland Publishing Co. 433 pp.

Pettijohn, F. J. 1949. *Sedimentary Rocks*. 3d ed. (1975). New York: Harper & Row. 736 pp.

Potter, P. E. 1963. *Late Paleozoic Sandstones of the Illinois Basin*. Urbana: Illinois State Geological Survey, Reports of Investigations no. 217. 92 pp.

Shelton, J. S. 1966. *Geology Illustrated*. San Francisco and London: W. H. Freeman. 434 pp.

Secondary Structures

Interesting secondary structures are found in most sediments, sedimentary rocks, and soils. They take peculiar shapes which stimulate the imagination. In a Geology Department there are probably more inquiries about such features than about minerals and fossils.

Secondary structures are formed by chemical action. Some result from solution, the more common from processes which segregate and concentrate deposits of minor constituents of the rock in which they are found. Some structures must have developed soon after the original deposition, others relate to movement of groundwater in the recent past. The chemical and physical environment of rock changes with burial and reexposure, the movement of groundwater, the covering vegetation, and so forth. The resulting alterations are not always specifically explainable, nor do all geologists subscribe to the same explanations.

Nodules are irregular masses of mineral matter generally different from the enclosing rock. They have rounded, rather smooth surfaces, often somewhat knobby. Chert nodules are abundant in the limestone units. When you find one, look along the same bedding plane where there may be others regularly spaced. They may even touch. Nodular cherts are generally gray to black with a white "cottony" rind (Fig. *171*). The composition of both is microcrystalline silica but the rind is granular. The white rind may be scraped off, but it is very hard and is so abrasive it will polish a knife blade and scratch glass.

Disks of marcasite or pyrite with a radiating crystalline structure have been found in some of the black shales above coal beds. These are much prized by mineral collectors.

Concretions are masses of material consisting of concentric layers generally surrounding a nucleus. They are more resistant than the enclosing rock, and their forms and relationships show that they developed after the original deposit. They are of many compositions, shapes, and sizes. Look for them in stream beds, old quarries, road cuts, and river banks. Concretions tend to be aligned along certain bedding planes.

Most famous of the concretions are the reddish elliptical masses, two to eight inches long, containing a carbonized plant fossil at the center (Fig. *172*). The fossils were deposited in muds accumulated in swamp waters low in oxygen. Subsequently, the chemistry changed and iron oxide was precipitated around the preserved organic matter, further protecting it.

171. Chert nodule from Silurian-age rocks near Thebes, Alexander County. The chert has a smooth texture, a procelain-like luster, and is hard but brittle. It is enclosed in a fine-grained limestone that has a more grainy texture and is easily scratched with a knife. Limestone effervesces on contact with a drop of dilute hydrochloric acid; chert does not.

Collection is best in old strip mines. Two localities now well-picked-over are near Cambria in Williamson County and Murphysboro in Jackson County. The buried plant-remains caused precipitation of iron oxide in the surrounding shale.

Other clay ironstones do not have an obvious organic center (Fig. *173*). These have more platy shells of iron carbonate or iron oxide. They are common in the Durry shale and in the Abbott Formation in southwestern Illinois. Occasionally concretions have tensional cracks, as if from dessiccation. When these are subsequently filled by another mineral, such as calcite, the name septaria is applied.

Other iron oxide or iron-manganese oxide concretions are found around plant rootlets in stratified sands and silts of the Pleistocene terraces and in many of the loess deposits.

Calcium carbonate concretions are found in the thick loess deposits near the Mississippi Valley bluffs from Jersey to Alexander counties. They range in color from cream to reddish brown. Most are hollow and have tension cracks. They take fascinating irregular and contorted shapes. Called "loess kindchen" or "children of the loess" by the Germans, they are the result of leaching of the upper part of the loess and precipitation at lower levels by soil-forming processes still active today.

Geodes are the particular favorite of mineral collectors. They are spheroidal masses with a

172. Iron oxide cemented concretion from shale stripped off the coal at Cambria, Williamson County. A carbon film of a fern of Pennsylvanian period is beautifully preserved.

173. Clay ironstone concretions enclosed in shale of the Abbott Formation southwest of Carbondale, Jackson County. The concretions are very hard to break when first exposed, but air and moisture causes them to spall off layer after layer.

resistant siliceous rind. Most are two to six inches in diameter but geodes more than a foot across are not uncommon. A hollow central cavity lined with crystals is the chief attraction (Fig. *174*). The outer rind is often botryoidal (having the form of a bunch of grapes) and may show concentric bands of chalcedony and quartz. Crystals face inward into the central cavity. Quartz and calcite are most common (see sec. Minerals.)

Geodes are found in impure limestones associated with chert and shaly layers especially in mid-Mississippian rocks of the central Mississippi Valley. The best collecting area in Southern Illinois is in the vicinity of Elsah in Jersey County.

The origin of geodes has been the subject of much study and speculation. One theory holds that a fossil with a hollow interior such as a crinoid or brachipod formed the locus of segregation of silica from the limestone. The silica replaced the carbonate of the shell and more was added to it. Movement of solution through the rind to form the crystalline lining of the cavity may have been the result of osmosis. The original hollow was presumably filled by sea water. Subsequent percolation of fresh water into the rock would produce the necessary pressure-differential to cause exchange through the rind.

Stylolites are seams with interlocking surfaces. Teeth-like projections of one side fit into

174. Quartz crystals line a geode cavity. The outer bumpy surface is said to have a botryoidal (grape-like) form. The two halves of this geode are mounted on a pair of bookends, four inches across.

sockets on the other (Fig. *175*). On the rock face in cross section they look much like an oscillogram of dark lines having an amplitude of one to three or four inches. Stylolites are found mainly in rather pure limestone, although they are known from sandstone.

The dark lines are of clay. At quarries look for blocks which have broken so as to expose the horizontal dimension. This may be nearly parallel to the bedding but does not represent a depositional surface. The tips of the projections generally are broken, but some idea of the extreme irregularity of the feature and of the clay accumulation can be gained.

Stylolites represent a residual accumulation of clay as the result of the solution process. Studies to estimate the amount of solution indicate that many feet of relatively pure limestone must have been dissolved in order to concentrate so much clay.

Stylolites are common in the limestone and marble used in public buildings, and are especially common in limestone slabs used on the walls of restrooms.

Veins originated long after the rock was solidified. Mineralizing solutions followed fault noticeable in quarries and other man-made excavations, but might be discovered in natural limestone cliffs and ledges.

In Southern Illinois veins are dominantly white clacite, but fluorspar is a common constituent. Quartz is not a vein mineral in our

175. This is a rare specimen of a stylolite in quartzose sandstone collected in Kentucky on the road from the ferry landing opposite Cave in Rock. Stylolites are very common, however, in rather pure sandstone.

region. The largest veins are in Hardin and Pope counties where they have been mined for fluorspar, but in most of the area veins are rarely more than a half inch thick.

Veins originated long after the rock was solidified. Mineralizing solutions followed fault and fracture openings, finally precipitating deposits when saturation took place. Veins extend directly across the bedding with a dominantly vertical component. In some cases veins intersect, or a succession of mineral depositions can be recognized where the later minerals cover or intersect earlier deposits.

Vugs are cavities filled with crystalline materials commonly associated with vein deposits and probably formed at the same time.

Things to do. Sedimentary rocks contain many peculiar features and structures. Take photographs and keep a careful record of the locations and the associations with othe features. Such a collection and its display would be an interesting project.

SUGGESTED READING

Pettijohn, F. J. 1949. *Sedimentary Rocks*. 3d ed. (1975). New York: Harper & Row. 736 pp.

Potter, P. E. 1963. *Late Palezoic Sandstones of the Illinois Basin*. Urbana: Illinois State Geological Survey, Reports of Investigations no. 217. 92 pp.

Shelton, J. S. 1966. *Geology Illustrated*. San Francisco and London: W. H. Freeman. 434 pp.

Structure of the Rock

Equality 1:62,500
Alto Pass 1:62,500
Herod 1:24,000

The most prominent features of sedimentary rocks are the near-horizontal bedding and the near-vertical fracture or joint pattern. The bedding is due to the original condition of deposition, reflecting changes in texture or composition. After the sediments were buried and compacted, differential stresses due to movements in the earth's crust caused systems of fractures to develop (Fig. *176*). Joints and bedding are zones of weakness along which weathering takes place and blocks are broken out (Fig. *177*).

Dip and Strike. The concepts of dip and strike are illustrated in Figure *178*. Dip represents the direction of the steepest slope of the rock layers, as shown by the arrow line on the diagram (see also Fig. *179*). Strike is at right angles to it, and is represented by the waterline between the pond and the bedding surface. Thus, the band of coal strip mines extending from Belleville to Harrisburg essentially follows the strike of sedimentary rocks making up the Illinois Basin. The surface of the ground is comparable to the pond surface in the diagram. The coal-bearing rock layers dip gradually into the basin, which attains its greatest depth in White County (Fig. *4*). The overburden rock is removed from the coal in strips successively deeper, following the coal down the dip. When stripping costs become too great underground methods must be used. In the up-dip direction from the strip mines there is no coal because it has been eroded away.

Folds. The regional structure of Illinois is a basin but the rock layers are wrinkled into local upfolds, "anticlines," and downfolds, "synclines." Such features of the bedrock are several miles across and not readily recognized even when traveling across them. In the plains provinces the bedrock is obscured by Pleistocene sediments, so the structural features had to be mapped by underground drilling and geophysical data.

The Eagle Creek syncline in southern Gallatin County is a textbook example of a syncline; it has coal beds preserved within it. See the structure map Figure *181* and the aerial photograph of the south flank where the topography and the rocks slope northward Figure *88*. Anticlines and domes are also of economic significance because they may be traps for petroleum.

Hicks Dome, shown on the aerial photograph (Fig. *180*), is a prominent feature in Hardin County. At the center, Devonian limestones form a dome-shaped hill. Surrounding the central hill is a lowland eroded into shales. A circular rim of hills composed of more resistant cherty limestones sets off the structure. Igneous dikes and explosive breccias (see sec. Igneous Rock) are associated with Hicks Dome. This structure was probably pushed upward by intruding molten magma. The main body did not reach the surface, and exploratory drill holes have not encountered any large igneous body.

Faults. Major fault systems have made profound impressions on the landscape of Southern Illinois. The map (Fig. *181*) identifies the location of faults many of which can be readily located on topographic maps. The actual rupture of the rock is difficult to see, although search in the indicated locations will reveal much fractured rock, slickensides (highly polished rock surfaces caused by friction), and mineralized fracture fillings.

Fault zones can be identified by elongate ridges or straight trends of valleys (Fig. *182*). Some fault lines cause high ridges or escarpments overlooking a lowland, very similar in appearance to a cuesta. The Shawneetown Fault is responsible for the uplift which caused

176. Sedimentary layering, or bedding, dipping slightly to the right. The beds are of different thicknesses and have a variety of detailed surface features. Two sets of near-vertical joints intersecting at right angles are well displayed at lower left. Kinkaid Lake spillway, Jackson County.

177. Blocks of sandstone have been removed by erosion during overflow of Kinkaid Lake along the spillway. The strong jointing of the rock makes removal easy.

178. Diagram representing dip and strike of sedimentary rocks.

179. A massive sandstone dips gently toward the upper right. Valley erosion has removed the rock to form a V upstream and in the direction of dip of the rock.

180. Looking northward at Hicks Dome, Hardin County. The central oval hill is of Devonian limestone surrounded by a grassed and cultivated lowland of more easily eroded shale. A rim of resistant cherty limestone hills is wooded. The valleys in the foreground radiate from the center. In the near distance is the Pennsylvanian sandstone cuesta and the Garden of the Gods. On the horizon the same rocks rise again on the north flank of Eagle Creek syncline to form Cave Hill and Gold Hill in Gallatin County.

181. Major fault systems of Southern Illinois.

Cave Hill to stand so high, overlooking the plains to the north. Here the vertical displacement is as much as 3,000 feet. The form of some displacements is vertical, but most of the larger displacements such as the Shawneetown Fault and Ste. Genevieve-Rattlesnake Ferry fault in Jackson-Union counties have had great lateral movements (Fig. *183*).

In Hardin County the faults served as passageways for movements of mineralizing solutions. They localized the mineral deposits, and several faults served as hosts for the deposits. Small displacements of a few feet may be found in many places (Fig. *184*). Small faults are encountered in most coal mines. Some are related to the large faults, but others seem to be only local features.

Unconformities

The succession of sedimentary layers generally seen in quarries and natural exposures are conformable, that is, deposits were laid down one after another without a significant time break. Rock units are essentially parallel or they wedge out, are cross-bedded or show similar evidence of near continuity. Unconformities show evidence of a significant break in deposition, with erosion of some units, a soil development, or widespread angular relation-

182. Looking north across the Shawnee Hills toward Harrisburg, Saline County. The ridge on the horizon at right is Cave Hill, a fault-line scarp. The straight valleys in foreground are also controlled by faults in the Lusk Creek area.

ship of the lower beds beneath a horizontal blanketing layer.

The most prominent unconformity is between the glacial till and the bedrock. Materials deposited perhaps 200,000 years ago rest on Paleozoic marine rocks formed over 250 million years ago, warped upward and subsequently weathered and eroded. The loess and glacial-lake deposits likewise rest unconformably on underlying materials.

Within the Paleozoic sequence the most widespread unconformity is at the base of Pennsylvanian rocks. Exposures of this contact are rarely seen, but a road cut on U.S. 51 northeast of Cobden, Union County, gives an excellent display of Pennsylvanian sandstone and basal conglomerate lying on shale and limestone of the Kinkaid Formation of Mississippian age.

Geologic Maps

A geologic map shows the distribution of rock units. The example in Figure *185* depicts a part of Hicks Dome in Hardin County. The sedimentary units have scalloped boundaries, making a "V" at the stream channels and curving up the hills (Fig. *178*). The "V's" indicate a west and northwestward dip of the rocks. The rocks in the center dip beneath the others, but the rocks appearing at the west have been eroded away from the center of the dome. Radial faults displace some units.

Things to do. Analogues to the joints and faults—folds too—can be found all around. Look for an old asphalt sidewalk, driveway, or parking lot, especially on a slope. Patterns of tension and shear cracks and compressional folds can usually be found. Concrete sidewalks next to a big tree have been broken by root growth. Wallpaper and plaster in an old house also display interesting structural analogues. A broken safety glass windshield is even more spectacular. The photograph of the wall of Johnston City Elementary School shows tension cracks due to settling toward the left (Fig. *186*).

183. Devil's Bake Oven near Grand Tower, Jackson County. The tilted and isolated limestone block stands within the complex of the Ste. Genevieve-Rattlesnake Ferry fault system.

SUGGESTED READING

Bond, D. C. 1971. *Background Materials for Symposium on Future Petroleum Potential of N.P.C. Region 9*. Urbana: Illinois State Geological Survey, Petroleum Series no. 96. 63 pp.

Heigold, P. C. 1968. *Notes on the Earthquake on November 9, 1968, in Southern Illinois*. Urbana: Illinois State Geological Survey, Environmental Geologic Notes no. 24. 16 pp.

Heyl, A. V., Jr., and Brock, M. R. 1961. *Structural Framework of the Illinois-Kentucky Mining District*. Washington, D.C.: United States Geological Survey, Professional Paper 424D, no. 294. pp. D3–D6.

Illinois State Geological Survey. Urbana. Many reports and maps on structure of coal and of oil fields.

Ross, C. A. 1963. *Structural Framework of Southernmost Illinois*. Urbana: Illinois State Geological Survey, Circular no. 351. 27 pp.

184. Small compressional fault exposed in a road cut on Illinois 146 near Dixon Springs State Park, Pope County. The left side has been pushed up and over that on the right. The structure is also visible on the other side of the road.

185. Part of the geologic map of the Herod Quadrangle taken from the Illinois Geological Survey Circular no. 413.

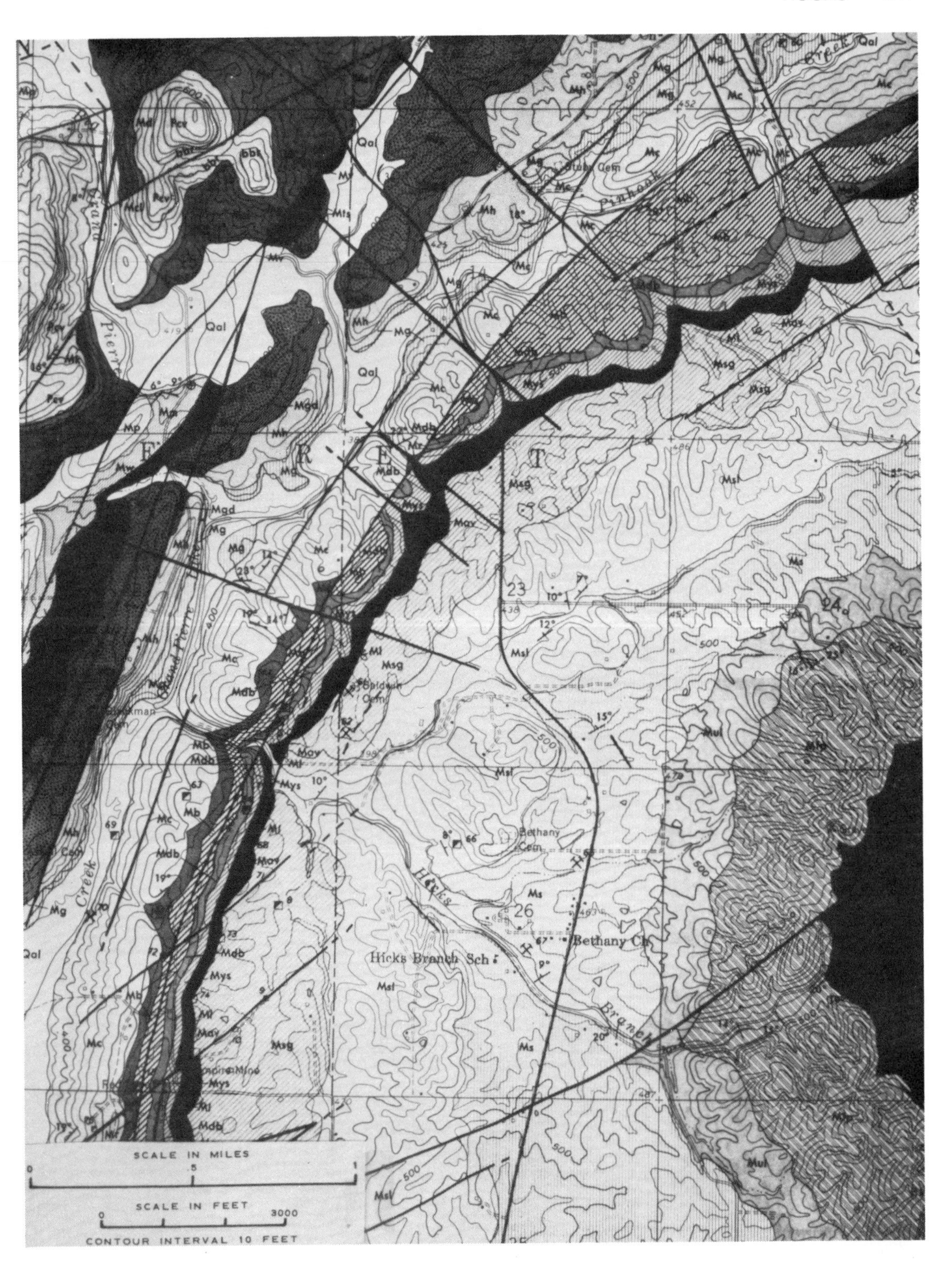
Hicks Branch Sch.
Bethany Ch
Bethany Cem.
Hicks
Branch
Grand Pierre
Creek
Pinhook
23
24
26
SCALE IN MILES
0
.5
1
SCALE IN FEET
0
3000
CONTOUR INTERVAL 10 FEET

186. Johnston City Elementary School, September 1973, before it was demolished. The cracks are due to tensional stresses caused by greater settling on the left side. This was brought about by roof collapse of an old coal mine and subsidence of the surface.

Minerals

Hardin County is famous among mineral collectors for beautiful crystals of fluorite and the minerals associated with it (Fig. *187*). Mining operations frequently break into openings lined with crystals up to several inches across. Specimens are displayed in museums all over the world.

Collecting firsthand is difficult today because few visitors are permitted in the mines. The old surface workings and accessible waste piles are well worked over. Good material is available for purchase at a number of rock shops in Cave in Rock, Rosiclare, and in the county.

187. The fluorite (fluorspar) crystals from Hardin County are famous the world over. The dark cubes are fluorite and the white crystals are calcite.

One might not expect to find many minerals in other parts of Southern Illinois but students in the mineralogy class at Southern Illinois University have found forty-three minerals in Jackson County alone; these are listed at the end of the section. The most favorable geologic conditions are in 1) the coal strip mines and old gob piles, 2) limestone quarries, 3) the glacial till, and 4) river or glacial gravels and sands. When collecting be sure to label the specimens, giving the location and the surrounding conditions.

Minerals of the spoil and gob piles are related to the coal deposits (Fig. *188*). A number of new minerals are actually formed after mining operations, developing on the newly exposed soil and tailings piles. Iron sulphide, either as pyrite or marcasite, is present in the coal and the black shale. On exposure to weathering it is rapidly oxidized and leached and recombines with weathering products of other minerals. Large "fish tail" crystals of gypsum are especially prized, while small crystals of white melanterite and yellow sulphur coat the surfaces of many rocks.

Be very cautious near gob and tailings ponds. They are unstable areas. Few gob piles are still on fire, but crusted surfaces may give way and acid wastes could burn the skin. Fine tailings may still be uncompacted and unable to support a person's weight.

Glacial till is especially well exposed in coal strip mines, but can be seen in many road cuts, eroding stream valleys, and on steep slopes of the Mt. Vernon Hill Country. The till contains crystalline igneous and metamorphic rocks brought from the Canadian Shield. Individual minerals can be identified in the coarser-grained rock.

Glacial gravels and sands also contain crystalline rocks. The terraces or the channel load of the Kaskaskia, Mississippi, Ohio rivers and the Wabash and its tributaries are a good source for these. Agates and japseroid cherts are prized finds.

The limestone quarries have veins of calcite and purple fluorite and in some places vuggy openings lined with crystals such as dolomite and sphalerite.

188. Efflorescences color the surface of a piece of oal exposed to air and moisture. Small veins of pyrite can be seen on the left. The light-colored portion consists largely of sulphates.

Geodes are common only in Jersey County near Elsah but are reported now and again from other localities (Fig. *169*). Geodes hold an especial fascination. They are dense and hard spherodial masses, commonly with a central cavity lined with crystals. The outer shell of the geodes is cherty, but may be banded with chalcedony and quartz. Many kinds of crystals are found inside, though quartz and calcite are most common. Perhaps the most startling filling is petroleum; its mode of origin is puzzling to geologists, too.

The following minerals have been collected in Jackson County, Illinois, by mineralogy classes.

1. Alum
2. Aragonite
3. Barite
4. Biotite
5. Calcite
6. Chalcopyrite
7. Chert
8. Chlorite
9. Dolomite
10. Epidote
11. Fluorite
12. Galena
13. Garnet
14. Goethite
15. Gypsum
16. Hematite
17. Hornblende
18. Ice
19. Kaolinite*
20. Kyanite
21. "Limonite"
22. Magnetite
23. Malachite
24. Marcasite
25. Melanterite
26. Muscovite
27. Olivine
28. Opal
29. Plagioclase
30. Pyrite
31. Pyrolusite
32. Pyroxene—probably augite
33. Quartz
34. Rozenite

35. Siderite
36. Sphalerite
37. Sulphur
38. Szomolnokite
39. Tourmaline
40. Vermiculite
41. Zircon
42. Illite*
43. Montmorillonite* (actually mixed layer illite-montmorillonite)

*Identification of clay minerals must be by X-ray diffraction.

Source: Courtesy of Dr. William Hood, Southern Illinois University-Carbondale, Geology Department.

Things to do. Mineral collecting is a hobby of millions of people. You can start your own by searching your home area, and gathering minerals on vacation trips. Be sure to label your specimens when you collect them. Later these may be classified systematically by area or even by economic use.

In our area the Southern Illinois Earth Science Club meets monthly in Mt. Vernon. They have speakers, exchange speciments, and arrange field trips and visits to one another. A Minerals Show is held once a year. Dealers from far and wide display their wares. The club welcomes new membres and visitors.

Several mineral collecting clubs are active in the St. Louis area; they also sponsor shows.

SUGGESTED READING

Baxter, J. W. et al. 1973. *A Geological Excursion to Fluorspar Mines in Hardin and Pope Counties, Illinois*. Urbana: Illinois State Geological Survey, Guidebook Series no. 11. 30 pp.

Biggs. D. L. 1957. *Petrography and Origin of Illinois Nodular Cherts*. Urbana: Illinois State Geology Survey, Circular no. 245. 25 pp.

Boegel, Hellmuth. 1967. *The Studio Handbook of Minerals*. Edited by John Sinkankas. New York: Viking Press (1971). 304 pp.

Fenton, C. L., and Zara, Louis. 1965. *Rocks, Minerals and Gems*. Prepared with the cooperation of the National Audubon Society. Garden City, N.Y.: Doubleday. 64 pp.

Illinois State Geological Survey. 1959. *Guide to Rocks and Minerals of Illinois*. Educational Series no. 5. Urbana. 40 pp.

Pearl, Richard M. 1956. *Rocks and Minerals*. New York: Barnes and Noble. 275 pp.

Sinotte, Stephen R. 1969. *The Fabulous Keokuk Geodes*, Vol. 1. Des Moines, Iowa: Wallace-Homestead. 292 pp.

Vanders, Iris, and Kerr, P. F. 1967. *Mineral Recognition*. New York: John Wiley & Sons. 316 pp.

Mineral Products

The mineral products used by man may be a source of interest and even detective skills. Stone has been used for construction purposes since pioneer days. Some limestone and sandstone were obtained locally, cut into dimension stone or rough stone for retaining walls, barn foundations, and so forth. Dimension stones have also been imported for sidewalks (still used in Cobden), for curbing (College St. in Carbondale), tombstones, millstones, and so forth.

Gravels and sands are as distinctive as rock. Their sources can be traced to Bald Knob, to the Metropolis upland, the Wabash terraces, and so on.

Also interesting but more difficult to trace are the manufactured products such as brick and tile. Some manufacturers stamped their names into the brick. Bricks from city streets and old buildings may reveal old industries long since defunct. The clays from the pits near Cobden and Olmsted have mostly been exported. Crushed rock from limestone quarries may have distinctive identifying features such as their fossils, oolitic texture, and chert. The mine waste from the fluorspar district has an abundance of white calcite rhombs and perhaps some unrecovered fluorspar or sphalerite.

Glass has also been manufactured in the past. Waste glass of several colors has been brought to the Southern Illinois University Geology Department for identification. It has the appearance of obsidian—often very attractive.

Fossils

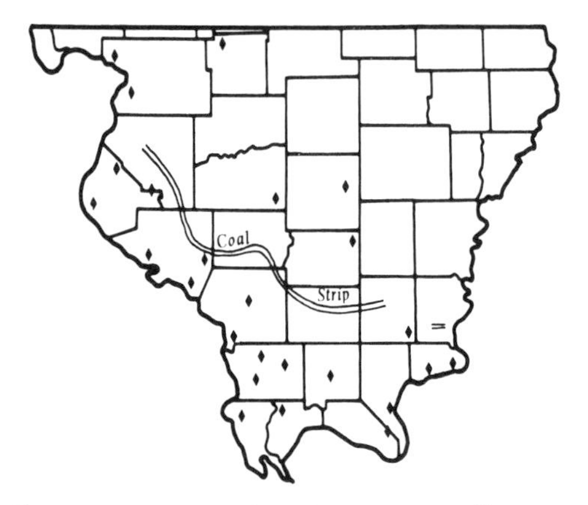

Fossils are abundant in many sedimentary rocks in Illinois. They are interesting and a challenge to collect. Collecting may be a casual hobby, just a desire to assemble an interesting variety of good specimen material, or a serious scientific effort to gather representative assemblages of fossil remains. The latter intent is aimed at reconstruction of the life and environment of the geologic interval represented by the rock unit.

Many outstanding collections are on display or available for study in the museums and university repositories of our area and at the State Geological Survey at Champaign, the United States Geological Survey, and the Smithsonian Institute in Washington, D.C.

Invertebrate fossils are most abundant. They are found in all natural divisions. The Chester Series limestones and calcareous shales are the richest (Figs. *189* and *190*), but some formations of the Valmeyer Series, also of Mississippian age, are as productive. These units are found in the southern Shawnee Hills and northern Ozark divisions. Similar fossils are found in the plains division. Shells of snails and ostracodes, small bivalves, are found in Pleistocene lake terrace sediments. Near the bluffs of the Mississippi River Valley the thick loess sections contain abundant snail shells. These were pulmonate woods dwellers rather than the aquatic forms which are found in the Pennsylvanian limestones and shales.

Plant fossils of ferns, seed ferns, scouring rushes, and segments of the trunks and roots of the great scale trees are found in Pennsylvanian and Chester rocks. The best collecting is on old spoil piles of strip mines where the hard ironstone nodules lie on the surface, washed free of finer material by erosion. Black carbon films and impressions of leaves are also found in gray shales but they are prone to disintegration. Impressions of trunks of scale trees are rather common in some of the sandstones (Fig. *191*) but these tend to lack detail. In coastal-plains clays, notably the Porter's Creek Clay of Paleocene age near Olmsted, Pulaski County, leaves of angiosperms are beautifully preserved. Much micro material requiring special techniques of separation is also present, but little detailed attention has been given to it.

Vertebrate remains are rare. Sharks' teeth and plates may be found in some of the Chester formations. Pleistocene alluvial and lake sediments have yielded some mammoth and other mammalian remains, but not so many as have been found in the western plains states.

Fossil Collecting

Fossils are found in most kinds of sediments and sedimentary rocks, so one should always keep his eyes open for them. However, they are most likely to be found in natural or man-made excavations where conditions favor erosion and weathering of the surface. Valleys where the stream flows across bedrock or undercuts the valley slopes are favorable natural sites, although massive sandstones are not good hosts. Quarries, road cuts, clay pits, strip mines are better, especially after a few years of exposure. The many small limestone P.W.A. quarries in the southern Shawnee Hills still yield fair collecting, but they are past their

189. A slab of fossiliferous limestone from the Golconda Formation, Union County. The limestone contains clay which speeds weathering and erosion so that the whole fossil eventually weathers out.

prime. Many highway road cuts offer excellent collecting, although caution is necessary because of the traffic. Some old spoil areas are readily accessible, but most require special permission. These, too, eventually become too weathered and stabilized. Several well-known collecting sites for plant fossils have been thoroughly picked over, and new concretions are exposed only slowly.

Fossil collectors should be thoughtful and not greedy. It is senseless to gather large numbers of fossils only to discard them or mix them unlabeled in a box with many others. Collected fossils should be labeled as to location, rock unit, date, and name of collector. Identification can be made with the help of reference books or with help from paleontologists at colleges and universities, and the State Geological Survey. Earth Science teachers in many of the area schools also can give assistance.

Collecting from limestone, especially clayey limestone, may be very rewarding. Look for specimens already loosened from the surface. Sometimes excellent specimens have been carried down into gullies and road ditches, although the source bed may be difficult to identify. Notice that different rock types have

190. A weathered surface with many loosened fossils. Golconda Formation, Union County.

191. A two-foot section of the bark of a scale tree showing the typical diagonal rows of diamond-shaped leaf scars. It is preserved in the Battery Rock sandstone of Pennsylvanian age at the spillway of Kinkaid Lake, Jackson County.

different assemblages; they represent different environments. Slabby rocks can often be split easily with hammer and chisel. Fossils embedded in massive and dense rocks probably cannot be removed successfully. Leave them untouched for the next searcher to see also. Too often one sees gouges around a fossil and so much damage that the remaining fossil is of little value.

You may organize your collection in a variety of ways, but geologic age, sedimentary environment, biologic classification, and geographic location are the four most important elements. Photographs of the collecting sites are very useful and a polaroid camera permits immediate notation. A map, carefully marked, will be of long-term value.

SUGGESTED READING

Collinson, C. W. 1959. *Guide for Beginning Fossil Hunters*. Urbana: Illinois State Geological Survey, Educational Series no. 4. 37 pp.

Fenton, C. L., and Fenton, M. A. 1958. *The Fossil Book*. Garden City, N.Y.: Doubleday.

Mathews, W. H., III. 1962. *Fossils*. New York: Barnes and Noble. 337 pp.

Seilacher, Adolph. 1967. Fossil Behavior. *Scientific American* 217: 72–80. (Offprint no. 872. San Francisco: W. H. Freeman.)

Appendixes

Index

Appendix A

Sources of Maps

Department of Transportation
Administration Building, Room 130
2300 South 31st Street
Springfield, IL 62764
(District Offices in Carbondale, East St. Louis, Effingham)

Up-to-date road and traffic maps for state, county, and city.

Illinois State Geological Survey
Natural Resources Building
Urbana, IL 61801

Topographic maps, also maps relating to geology, mineral and water resources. Write for list of publications and Field Trip Guide Leaflets.

Illinois Historic Sites Survey
1202 W. Main Street
Carbondale, IL 62901

County history maps.

Illinois State Water Survey
P. O. Box 232
Urbana, IL 61801

Maps of some lakes and rivers.

Office of Forest Supervisor
Shawnee National Forest
317 E. Poplar Street
Harrisburg, IL 62946

Topographic maps (by township), also Shawnee Forest map showing roads and recreation areas.

Outdoor Illinois
320 S. Main Street
Benton, IL 62812

Historic and scenic maps.

Rockford Map Publishers, Inc.
4525 Forest View Avenue
P. O. Box 6126
Rockford, IL 61125

County land-ownership plat books.

United States Department of Agriculture
Soil Conservation Service
(At most county seats; listed in telephone book.)

County and local soils maps; aerial photographs for inspection.

United States Geological Survey
Distribution Section
1200 S. Eads Street
Arlington, VA 22202

Topographic maps ($1.25 each). Write for Index map.

Map Libraries

Southern Illinois University
Carbondale, IL 62901

University of Illinois
Urbana, IL 61801

Topographic maps and aerial photographs.

Aerial Photographs and Satellite Imagery

Agriculture Stabilization and Conservation Service
Western Aerial Photographic Laboratory
A.S.C.S.-U.S.D.A.
2505 Parley's Way
Salt Lake City, UT 84109

Aerial photographs showing large-scale stereo coverage in sets of counties. Repeated flights from 1938, about a decade apart. Individual photographs may be located on an index photograph.

United States Department of Interior
Geological Survey
Earth Resources Observation Satellite (EROS) Data Center
Sioux Falls, SD 57198

High-level imagery (such as Fig. 21). Skylab, and other photographs. Request listing of coverage of specific area.

Appendix B

Federal and State Park Agencies

Illinois Department of Conservation
Division of Education
State Office Building
Springfield, IL 62706

Maps and folders for parks and nature preserves.

United States Army Corps of Engineers
St. Louis District
210 N. 12th Street
St. Louis, MO 63101

Or

Ohio River Division
550 Main Street
P. O. Box 1159
Cincinnati, OH 45202

River navigation charts, folders for reservoir and recreation areas.

United States Department of Agriculture
Soil Conservation Service
(At most county seats; listed in telephone book.)

Information about conservancy district, aerial photographs, soils maps, pamphlets, and so forth.

United States Department of Interior
Fish and Wildlife Service
Crab Orchard National Wildlife Refuge
P. O. Box J
Carterville, IL 62918

Maps of refuge area.

United States Forest Service
Shawnee National Forest Headquarters
Harrisburg, IL 62946

Ranger stations at Elizabethtown, Jonesboro, Murphysboro, and Vienna.

Brochures for camping and scenic areas. Area maps, topographic quadrangles of forest area by township.

Appendix C

Stratigraphy

Geologists have applied names to the rock formations of our area. Formations are intended to be rock units identifiable by distinctive characteristics of lithology, such as texture, composition, sedimentary structures, and fossils. They are named from a geographic locality where they are exposed at the surface of the ground. In Illinois the formations at the bottom of the column tend to be more uniform than the Chesterian and Pennsylvanian rocks, which were deposited where the local environmental conditions were rapidly fluctuating. Classification is more difficult and formation units generally contain more than one rock type. In some places distinctive layers may be given "member" names, for example, the Greenup limestone in the Mattoon Formation.

Rocks were, of course, deposited during a certain interval of time. However, while we conceive of time as continuous, there may have been long intervals when no rocks were being formed in a given locality. In the left column of the following chart the names represent the geologic time periods (and also the rocks which were formed during that time interval). In the second column the radiometric age in millions of years is shown at the boundaries between periods. The epoch series are time-rock subdivisions of the period systems and may include several formations.

The more prominent rock type is given for each unit, although many formations include several lithologies. Thicknesses are quite variable because of the large area and variable conditions whithin the Illinois Basin.

The geologic map of Illinois (Fig. 2) shows the surface distribution of the major units. The formations are described in detail in many reports of the Illinois Geological Survey (for example, H. B. Willman et al. 1975. *Handbook of Illinois Stratigraphy*. Bulletin no. 95. 261 pp.), and especially those relating to local areas.

STRATIGRAPHIC UNITS IN SOUTHERN ILLINOIS

SYSTEM PERIODS	AGE MIL. YRS.	EPOCH–SERIES	ORIGIN AND CHARACTER	FORMATION MEMBER	ROCK TYPE	THICKNESS FEET
		Holocene	Soil, river deposits			
QUATERNARY		Pleistocene Wisconsinan Glacial Sangamonian Interglacial Illinoian Glacial Yarmouthian Interglacial Kansan Glacial	Continental–glacial, river, lake and swamp deposits, soils			0-300
		––EARLY RECORD MISSING––				
	1.5					
		Pliocene	Continental--gravel, sand	Mounds		0-50
		––OLIGOCENE AND MIOCENE SERIES MISSING IN ILLINOIS––				
TERTIARY		Eocene	Deltaic sand	Wilcox		0-300
		Paleocene	Marine clay and silt	Porters Creek Clayton		75-100 5-10
	65					
CRETACEOUS		Gulfian	Deltaic and marine clay sand and gravel	Owl Creek McNairy Tuscaloosa		0-10 125-150 0-20
	125					

SYSTEM PERIODS	AGE MIL. YRS.	EPOCH—SERIES	ORIGIN AND CHARACTER	FORMATION MEMBER	ROCK TYPE	THICKNESS FEET
		—PERMIAN TRIASSIC AND JURASSIC MISSING IN ILLINOIS—				
	280					
PENNSYLVANIAN		Virgilian	Mainly marine shale with limestone, sandstone and coal	Mattoon		700±
				Greenup	ls	
				Shumway	ls	
		Missourian		Bond		100-300
				Millersville	ls	
				Shoal Creek		
				Modesto		200-500
				No. 8	coal	
		Desmoinesian	Marine, continental deltaic shale and sandstone with the principal commercial coals, some limestone in upper part	Piasa	ls	
				Carbondale		200-400
				No. 7	coal	
				No. 6 Herrin	coal	
				No. 5 Springfield or Harrisburg	coal	
				No. 2 Murphysboro	coal	
				Spoon		10-350
				Dekoven	coal	
				Davis	coal	
				Curlew	ls	
		Atokan	Continental and deltaic sandstone, siltstone and shale; conglomeratic zones in lower part	Abbott		0-350
				Murray Bluff	ss	
				Grindstaff	ss	
		Morrowan		Caseyville		0-600
				Pounds	ss	
				Drury	shale	
				Battery Rock	ss	
	315					
MISSISSIPPIAN		Chesterian	Marine, deltaic cyclical deposits of sandstone, limestone, shale	Grove Church	sh	0-1400
				Kinkaid	ls	
				Degonia	ss	
				Clore	ls, sh.	
				Palestine	ss	
				Menard	ls, sh	
				Waltersburg	ss	
				Vienna	ls	
				Tar Springs	ss	
				Glen Dean	ls	
				Hardinsburg	ss	
				Golconda Group	ls, sh	
				Cypress	ss	
				Ridenower	ls	
				Bethel	ss	
				Downeys Bluff	ls	
				Renault	ls	
		Valmeyeran	Marine—dominantly limestone, some cherty	Aux Vases	ss	10-80
				Ste. Genevieve		60-200
				"McCloskey"	oolitic	
				St. Louis	chty, ls	1500-500
				Salem	ls	100-400
				Ullin	ls	70-800

SYSTEM PERIODS	AGE MIL. YRS.	EPOCH–SERIES	ORIGIN AND CHARACTER	FORMATION MEMBER	ROCK TYPE	THICKNESS FEET
				Fort Payne	cht	0-500
			Deltaic in south and east	Borden-Springville	sh	10-700
		Kinderhookian	Marine shale and limestone	Chouteau	ls	
				Hannibal	sls	0-100
				Saverton	sls	
	345					
DEVONIAN		Upper	Black shale with spore carps	Grassy Creek		5-300
		Middle	Marine limestone	Alto-Blocher	ls, sh	0-100
			Sandstone at base	Lingle	ls	20-120
				Grand Tower	ls	0-200
				Dutch Creek	ss	0-15
		Lower	Marine limestone, very cherty or siliceous	Clear Creek	cht	
				Backbone	ls	30-150
				Grassy Knob	cht	120-250
		"Hunton" used for Devonian-Silurian Limestone Section		Bailey	ls	250-450
	395					
SILURIAN		Cayugan	Marine limestone, some shaly, local reefs	Moccasin Springs	ls, sh	100-800
		Niagaran		St. Clair	ls	20-100
		Alexandrian		Sexton Creek	chty, ls	15-80
				Edgewood	ls	0-60
				Girardeau	ls	0-40
	435					
ORDOVICIAN		Cincinnatian	Marine shale, siltstone limestone and dolomite	Orchard Creek		
				Thebes		0-100
				Cape	0-20	0-20
		Champlainian	Marine limestone	Kimmswick	ls, chty	
		"Trenton"				600
		NO SURFACE EXPOSURES BELOW HERE				
			Marine fossiliferous limestone and shale; sandstone at base	Decorah Group	sh, ls	
				Plattin	ls	
				Joachim	ls	
				Dutchtown	ls	
				St. Peter	ss	
		Canadian Knox Megagroup extends down into Cambrian	Marine dolomite, cherty and sandy; sandstone	Jefferson City	dol, cht	
				Roubidoux	ss, dol	17003
				Gasconade	chty, dol	
	500					
CAMBRIAN		Croixan	Marine sandstone shale and dolomite	Trempealeau		
				Eau Claire		700±
				Mt. Simon		500±
	600					
PRECAMBRIAN		Crystalline basement, apparently mostly granitic similar to St. Francis Mountains of Missouri				

Index

The numbers set in boldface type indicate a map, a photograph, or a diagram.